T0337827

CHIPLESS RADIO FREQUENCY IDENTIFICATION READER SIGNAL PROCESSING

CHIPLESS RADIO FREQUENCY IDENTIFICATION READER SIGNAL PROCESSING

NEMAI CHANDRA KARMAKAR, PHD
PRASANNA KALANSURIYA, PHD
RUBAYET E. AZIM
RANDIKA KOSWATTA, PHD

For general information on our other products and services or for technical support, please contact our Customer Care Department within the United States at (800) 762-2974, outside the United States at (317) 572-3993 or fax (317) 572-4002.

Wiley also publishes its books in a variety of electronic formats. Some content that appears in print may not be available in electronic formats. For more information about Wiley products, visit our web site at www.wiley.com.

Library of Congress Cataloging-in-Publication Data:

Names: Karmakar, Nemai Chandra, 1963– author.
Title: Chipless radio frequency identification reader signal processing /
    Nemai Chandra Karmakar.
Description: Hoboken, New Jersey : John Wiley and Sons, Inc., [2016] |
    Includes bibliographical references and index.
Identifiers: LCCN 2015024764 (print) | LCCN 2015027791 (ebook) |
    ISBN 9781119215752 (cloth) | ISBN 9781119215646 (pdf) | ISBN 9781119215776 (epub)
Subjects: LCSH: Radio frequency identification systems.
Classification: LCC TK6570.I34 K365 2016 (print) | LCC TK6570.I34 (ebook) |
    DDC 621.3841/92–dc23
LC record available at http://lccn.loc.gov/2015024764

Set in 11/13pt TimesTen by SPi Global, Pondicherry, India

Printed in the United States of America

CONTENTS

PREFACE

Introduction to Radio Frequency Identification (RFID): RFID is a wireless modulation and demodulation technique for automatic identification of objects, tracking goods, smart logistics, and access control. RFID is a contactless, usually short-distance transmission and reception technique for unique ID data transfer from a tagged object to an interrogator (reader). The generic configuration of an RFID system comprises (i) an ID data-carrying tag, (ii) a reader, (iii) a middleware, and (iv) an enterprise application. The reader interrogates the tag with the RF signal, and the tag in return responds with an ID signal. Middleware controls the reader and processes the signal and finally feeds into enterprise application software in the IT layer for further processing. The RFID technology has the potential of replacing barcodes due to its large information-carrying capacity, flexibility in operation, and versatility of application [1]. Due to its unique identification, tracing, and tracking capabilities, RFID also gives value-added services incorporating various sensors for real-time monitoring of assets and public installations in many fields. However, the penetration of RFID technology is hindered due to its high price tag, and many ambitious projects have stalled due to the cost of the chips in the tags. Chipless RFID tags mitigate the cost issues and have the potential to penetrate mass markets for low-cost item tagging [2]. Due to its cost advantages and unique features, chipless RFID will dominate 60% of the total RFID market with a market value of $4 billion by 2019 [3]. Since the removal of the microchip causes a chipless tag to have no intelligence-processing capability, the signal processing is done only in the reader. Consequently, a fully new set of requirements and challenges need to be incorporated and addressed, respectively, in a chipless RFID tag reader. This book addresses the new reader architecture and signal processing techniques for reading various chipless RFID tags.

Recent Development of Chipless RFID Tags: IDTechEx (2009) [3] predicts that 60% of the total tag market will be occupied by the chipless tag if the tag can be made at a cost of less than a cent. However, removal of an application-specific integrated circuit (ASIC) from the tag is not a trivial task as it performs many RF signal and information-processing tasks. Intensive investigation and investment are required for the design of low cost but robust passive microwave circuits and antennas using low-conductivity ink on low-cost and lossy substrates. Some types of chipless RFID tags are made of microwave resonant structures using conductive ink. Obtaining high-fidelity (high-quality factor) responses from microwave passive circuits made of low-conductivity ink on low-cost and lossy materials is very difficult [4]. Great design flexibility is required to meet the benchmark of reliable and high-fidelity performance from these low-grade laminates and poor conductivity ink. This exercise has opened up a new discipline in microwave printing electronics in low-grade laminates [5].

The low-cost chipless tag will place new demands on the reader as new fields of applications open up. IDTechEx [3] predicts that, while optical barcodes are printed in only a few billions a year, close to one trillion (>700 billion) chipless RFIDs will be printed each year. The chipless RFID has unique features and much wider ranges of applications compared to optical barcodes. However, very little progress has been achieved in the development of the chipless RFID tag reader and its control software, because conventional methods of reading RFID tags are not implementable in chipless RFID tags. As for an example, handshaking protocol cannot be implemented in chipless RFID tags. Dedicated chipless RFID tag readers and middleware [6] need to be developed to read these tags reliably.

The development of chipless RFID has reached its second generation with more data capacity, reliability, and compliance to some existing standards. For example, RF-SAW tags have new standards, can be made smaller with higher data capacity, and currently are being sold in millions [7]. Approximately 30 companies have been developing TFTC tags. TFTC tags target the HF (13.65 MHz) frequency band (60% of existing RFID market) and have read–write capability [7]. However, neither RF-SAW nor TFTC is printable and could not reach sub-cent-level prices. In generation-1 of conductive ink-based fully printable chipless RFID tag development, few chipless RFID tags, which are in the inception stage, have been reported in the open literature. They include a capacitive gap-coupled dipole array [8], a reactively loaded transmission line [9], a ladder network [10], and finally a piano and a Hilbert curve fractal resonators [11]. These tags are in prototype stage, and no further development to commercial grade has been reported to date. A comprehensive review of chipless RFID can be found in the author's most recent books [12].

Following 20 years of RF and microwave research experiences, the author has pioneered multi-bit chipless RFID research [13, 14]. For the last 10 years at Monash University, the author's research activities include numerous chipless tag and reader developments as follows.

At Monash University, the author's research group has developed a number of printable, multi-bit chipless tags featuring high data capacity. These tags can be categorized into two types: *retransmission based* and *backscattered based*. The *retransmission-based* tag, originally presented by Preradovic et al. [13], uses two orthogonally polarized wideband monopole antennas and a series of spiral resonators. The RFID reader sends a UWB signal to the tag, and the receiving antenna of the tag receives it, and then it passes through the microstrip transmission line. The gap-coupled spiral resonator-based stopband filters create attenuations and phase jumps in designated frequency bands, and this magnitude and phase-encoded signal is retransmitted back to the reader by the tag's transmitting antenna. The attenuation in the received signal due to the resonator encodes the data bits. In two Australian Research Council (ARC) Grants (DP0665523: *Chipless RFID for Barcode Replacement*, 2006–2008, and LP0989652: *Printable Multi-Bit Radio Frequency Identification for Banknotes*, 2009–2011), the author developed up to 128 data bits of chipless RFID with four slot-loaded monopole antennas and wideband feed networks [15]. This chipless tag is fully printable on polymer substrate.

Backscatter-Based Chipless Tag: Balbin et al. [13] have presented a multiantenna backscattered chipless tag. Here, only the resonator structure is present on the tag, and as no transmitter–receiver tag antenna exists, it is more compact than retransmission-type tags. The interrogation signal from a reader is backscattered by the tag. By analyzing this backscattered signal's attenuation at particular frequencies, the tag ID is decoded.

Monash University Chipless RFID Systems: Under various research grant schemes, the CI has developed various chipless RFID tag reader architectures and associated signal processing schemes. To date, four different varieties of chipless RFID tag readers have been developed for the 2.45, 4–8, and 22–26.5 GHz frequency bands. Feasibility studies of advanced level detection [13] and error correction algorithm have been developed.

As stated [2, 12–14], the author's group has developed four different varieties of chipless RFID tag readers in various frequency bands at 2.45, 4–8, and 22–26.5 GHz frequency bands. The readers comprised an RF transceiver section, a digital control section, and a middleware (control and processing). The reader transmitter comprises a swept frequency voltage-controlled oscillator (VCO) [6, 16]. The VCO is controlled by a tuning voltage that is generated by a digital-to-analog converter (DAC). Each frequency over the ultra-wideband (UWB) from 4 to 8 GHz is generated with a single tuning voltage from the DAC. In addition, the VCO has a finite settling time to generate a CW signal against its tuning voltage. Combining all these operational requirements and linearity of the devices, the UWB signal generation is a slow process (taking a few seconds to read a tag). To alleviate this problem and improve the reading speed, some corrective measures can be taken. They are (i) high-speed ADC and (ii) low settling time VCO. These two devices will be extremely expensive if available in the market. The reader cost will be very high to cater for the requirement specifications of the chipless RFID reader. In this regard, signal processing for advanced detection techniques alleviates the reading process in greater details. Also, the sensitivity of the reader architecture using dual antenna in bistatic radar configuration and I/Q modulation techniques can be greatly enhanced. Highly sensitive receiver design is imperative in detecting very weak backscattering signal from the chipless tag. With the presence of interferers and movement and the variable trajectory of the moving tags, this situation is worsened. In this regard, a highly sensitive UWB reader receiver needs to be designed. Designing such a receiver is not a trivial task where the power transmission is limited by UWB regulations. I\Q modulation in the receiver will improve the sensitivity to a greater magnitude.

Additional to this high-sensitivity receiver design, high-end digital board with a powerful algorithm will alleviate the reading process. The digital board serves as the centerpiece of the reader where data would be processed, and numerous control signals to the RF section of the reader would be generated. The digital board has a Joint Test Action Group (JTAG) port where a host PC can be connected to monitor, control, and reprogram the reader if necessary. In addition, it is also the host to the power supply circuit, which is used to generate the necessary supply voltages for most components of the reader. The digital board consists of (i) an FPGA board with ADC, (ii) a power supply circuit, and (iii) a DAC. High sampling rate A/D and D/A converters and an FPGA controller will augment the powerful capturing and processing of backscattering signals. The digital electronics and interface with a PC will accommodate custom-made powerful algorithm such as singular value decomposition for improved detection [17] and time–frequency analysis [18] for localization [19] and anticollision [20] of chipless RFID tags. All this control algorithm and signal processing software will be innovations in the field. The book has addressed these advanced level analog and digital designs of the chipless RFID reader.

In conventional chipped RFID system, established protocols are readily available for tag detection and collision avoidance. Reading hundreds of proximity tags with the flick of an eye is commonplace. However, reading multiple chipless RFID tags in close proximity is not demonstrated as yet. RFID middleware is an IT layer to process the captured data from a tag by a reader. Middleware applies filtering, formatting, or logic to tag data captured by a reader so that the data can be processed by a software application. For chipped RFID, there are established protocols for these tasks. However, in chipless RFID, tasks such as handshaking are not possible. Therefore, a completely new set of IT layers needs to be developed. Raw data obtained from a chipless tag need to be processed and denoised, and new techniques need to be developed. They are as follows: (i) signal space representation of chipless RFID signatures [21], (ii) detection of frequency signature-based chipless RFID using UWB impulse radio interrogation [22], (iii) a singularity expansion method for data extraction from chipless RFID [23], and finally (iv) noise reduction and filtering techniques [23, 24]. These methods will improve the efficacy and throughput of various types of reading processes. For example, in (i), tag signatures are visualized as signal points in a signal space (Euclidian space). (i) performs better than a threshold-based approach to detection. In (ii), the received signal from a chipless tag is processed in time domain, and information-carrying antenna mode RCS is processed to identify tags. In (iii), transient response from the tag is processed in poles and residues, and tag ID is detected. In (iv), wavelet transforms and prolate spheroidal wave functions are used for noise filtering. All these detection and filtering techniques are investigated in the context of the chipless RFID system, and the best approach to tag detection is integrated in the IT application layer.

The book aims to provide the reader with comprehensive information with the recent development of chipless RFID signal processing, software development algorithm, and protocols. To serve the goal of the book, the book features ten chapters in two sections. They offer in-depth descriptions of terminologies and concepts relevant to chipless RFID detection algorithm and anticollision protocols related to the chipless RFID reader system. The chapters of the book are organized into two distinct topics: (i) *Section 1:* Detection and Denoising and (ii) *Section 2:* Multiple Access and Localization. In chapter 1 chipless RFID fundamentals with a comprehensive overview are given. The physical layer development of reader architecture for conventional RFID systems is an established discipline. However, a physical layer implementation of the chipless RFID reader is a fully new domain of research. This author group has already published a book in this area [25]. This book focuses on the back-end postprocessing and detection algorithms for chipless RFID reader. Various detection algorithms for chipless RFID tags such as signal space representation, time-domain analysis, singularity expansion method for data extraction, and finally denoising and filtering techniques for frequency signature-based chipless RFID tags are presented in Chapters 2–5. Collision and error correction protocols, multi-tag identification through time–frequency analysis, FMCW-radar-based collision detection and multi-access for chipless RFID tags, and localization and tracking of tag are presented in Chapters 6–9. Finally, a state-of-the-art chipless RFID tag reader that incorporates all the physical and IT layer developments stated previously are presented in Chapter 10. The chapter has demonstrated how the reader can mitigate interferences and collisions keeping the data integrity in reading multiple tags in challenging environments such as retails, libraries, and warehouses.

In the book, utmost care has been paid to keep the sequential flow of information related to the chipless RFID reader architecture and signal processing. Hope that the book will serve as a good reference of chipless RFID systems and will pave the ways for further motivation and research in the field.

REFERENCES

1. K. Finkenzeller, *RFID Handbook: Fundamentals and Applications in Contactless Smart Cards, Radio Frequency Identification and Near-Field Communication*, 3rd Revised edition. Hoboken: John Wiley & Sons, Inc., 2010.
2. S. Preradovic and N. C. Karmakar, "Chipless RFID: bar code of the future," *IEEE Microwave Magazine*, vol. 11, pp. 87–97, Dec 2010.

3. IDTechEx (2009). *Printed and Chipless RFID Forecasts, Technologies and Players 2009–2029.*

4. R. E. Azim, N. C. Karmakar, S. M. Roy, R. Yerramilli, and G. Swiegers, "Printed Chipless RFID Tags for Flexible Low-cost Substrates," in: *Chipless and Conventional Radio Frequency Identification: Systems for Ubiquitous Tagging.* Hoboken: IGI Global, 2012

5. R. Yerramilli, G. Power, S. M. Roy, and N. C. Karmakar, "*Gravure Printing and Its Application to RFID Tag Development,*" *Proceedings of the Materials Research Society Fall Meeting 2011*, Boston, USA, November 28–December 2, 2011.

6. S. Preradovic and N. C. Karmakar, "*Multiresonator Based Chipless RFID Tag and Dedicated RFID Reader,*" *Digest 2010 IMS*, Anaheim, California, USA, May 23–28, 2010.

7. IDTechEx, *RFID Forecasts, Players and Opportunities 2009–2019, Executive Summary*, 2009.

8. I. Jalaly and I. D. Robertson, "*Capacitively-Tuned Split Microstrip Resonators for RFID Barcodes,*" in *European Microwave Conference EuMC, 2005,* Paris, France, October 4–6, 2005, p. 4.

9. L. Zhan, H. Tenhunen, and L.R. Zheng, "*An Innovative Fully Printable RFID Technology Based on High Speed Time-Domain Reflections,*" *Conference on High Density Microsystem Design and Packaging and Component Failure Analysis, 2006. HDP '06*, Stockholm, Sweden, June 27, 2006, pp. 166–170.

10. S. Mukherjee, "*Chipless Radio Frequency Identification by Remote Measurement of Complex Impedance,*" in *European Conference on Wireless Technologies, 2007*, Munich, Germany, 2007, pp. 249–252.

11. J. McVay, et al., "*Space-Filling Curve RFID Tags,*" in *IEEE Radio and Wireless Symposium, 2006*, San Diego, January 17–19, 2006, pp. 199–202.

12. S. Preradovic and N. C. Karmakar, *Multiresonator-Based Chipless RFID: Barcode of Future.* New York: Springer, 2012.

13. S. Preradovic, I. Balbin, N. C. Karmakar, and G. F. Swiegers, "Multiresonator-based chipless RFID system for low-cost item tracking," *IEEE Transactions on Microwave Theory and Techniques*, vol. 57, Issue 5, Part 2, 2009, pp. 1411–1419.

14. I. Balbin and N. C. Karmakar, "Multi-Antenna Backscattered Chipless RFID Design," in: *Handbook of Smart Antennas for RFID Systems*, Wiley Microwave and Optical Engineering Series. Hoboken: John Wiley & Sons, Inc., pp. 415–444, 2010.

15. I. Balbin, *Chipless RFID Transponder Design*, PhD Dissertation, Monash University, 2010.

16. R. V. Koswatta and N. C. Karmakar, "A novel reader architecture based on UWB chirp signal interrogation for multiresonator-based chipless RFID tag reading," *IEEE Transactions on Microwave Theory and Techniques*, vol. 60, no. 9, pp. 2925–2933, 2012.

17. A. T. Blischak and M. Manteghi, "Embedded singularity chipless RFID tags," *IEEE Transactions on Antennas and Propagation*, vol. 59, pp. 3961–3968, 2011.

18. B. Boashash Editor (2003). *Time Frequency Signal Analysis and Processing, A Comprehensive Reference*. Oxford: Elsevier, 2003.

19. Anee, R. and Karmakar, N. C., "Chipless RFID tag localization," *IEEE Transactions on Microwave Theory and Techniques*, vol. 61, no. 11, pp. 4008–4017, 2013.

20. R. Azim and N. Karmakar, "*A Collision Avoidance Methodology for Chipless RFID Tags*," *Proceedings of the 2011 Asia Pacific Microwave Conference*, Melbourne, Australia, December 5–8, 2011, pp. 1514–1517.

21. P. Kalansuriya, N. C. Karmakar and E. Viterbo, "*Signal Space Representation of Chipless RFID Tag Frequency Signatures*," *Proceedings of the 2011 IEEE Global Telecommunications Conference (GLOBECOM 2011) IEEE GLOBECOM 2011*. Houston, TX, December 5, 2011.

22. P. Kalansuriya, N. C. Karmakar and E. Viterbo, "On the detection of frequency-spectra based chipless RFID using UWB impulsed interrogation," *IEEE Transactions on Microwave Theory and Techniques*, vol. 60, no. 12, pp. 4187–4197, 2012.

23. M. Manteghi, "*A Novel Approach to Improve Noise Reduction in the Matrix Pencil Algorithm for Chipless RFID Tag Detection*," *Digest 2010 IEEE Antennas and Propagation Society International Symposium (APSURSI), IEEE APSURSI2010*, Toronto, ON, July 11–17, 2010, pp. 1–4.

24. A. Lazaro, A. Ramos, D. Girbau, and R. Villarino, "Chipless UWB RFID tag detection using continuous wavelet transform," *IEEE Antennas and Wireless Propagation Letters*, vol. 10, pp. 520–523, 2011.

25. N. C. Karmakar, R. V. Koswatta, P. Kalansuriya, and R. Azim, *Chipless RFID Reader Architecture*, Boston: Artech House, 2013.

CHAPTER 1

INTRODUCTION

1.1 CHIPLESS RFID

Radio frequency identification (RFID) is a wireless data communication technology widely used in various aspects in identification and tracking. In this era of communication, information, and technology, RFID is undergoing tremendous research and developments. It has the potential of replacing barcodes due to its information capacity, flexibility, reliability, and versatilities in application [1]. The unique identification, tracking, and tracing capabilities of RFID systems have the potential to be used in various fields like real-time asset monitoring, tracking of item and animals, and in sensor environments. However, the mass application of RFID is hindered due to its high price tag, and many ambitious projects had been killed due to the cost of chipped tags. The low-cost alternative of chipped RFID system is the printable chipless RFID that has the potential to penetrate mass markets for low-cost item tagging [2]. The chipless tag doesn't have any chips, and hence, the most burdens for signal and data processing go to the reader side. This introduces a set of new challenges and requirements for the chipless RFID reader that need to be addressed. This book comprises the new

Chipless Radio Frequency Identification Reader Signal Processing, First Edition.
Nemai Chandra Karmakar, Prasanna Kalansuriya, Rubayet E. Azim and Randika Koswatta.
© 2016 John Wiley & Sons, Inc. Published 2016 by John Wiley & Sons, Inc.

advanced signal processing and tag detection methods that are being used in chipless RFID for identification and tracking of tags.

RFID is an evolving wireless technology for automatic identifications, access controls, asset tracking, security and surveillance, database management, inventory control, and logistics. A generic RFID system has two main components: a tag and a reader [3]. As shown in Figure 1.1, the reader sends an interrogating radio frequency (RF) signal to the tag. The interrogation signal comprises clock signal, data, and energy. In return, the tag responds with a unique identification code (data) to the reader. The reader processes the returned signal from the tag into a meaningful identification code. Some tags coupled with sensors can also provide data on surrounding environment such as temperature, pressure, moisture contents, acceleration, and location. The tags are classified into active, semi-active and passive tags based on their onboard power supplies. An active tag contains an onboard battery to energize the processing chip and to amplify signals. A semi-active tag also contains a battery, but the battery is used only to energize the chip, hence yields better longevity compared to an active tag. A passive tag does not have a battery. It scavenges power for its processing chip from the interrogating signal emitted by a reader; hence, it lasts forever. However, the processing power and reading distance are limited by the transmitted power (energy) of the reader. The middleware does the back-end processing, command, and control and interfacing with enterprise application as shown in Figure 1.1.

As mentioned previously, the main hindrance in mass deployment of RFID tags for low-cost item tagging is the cost of the tag. The cost of the tag mainly comes from the application-specific integrated circuit

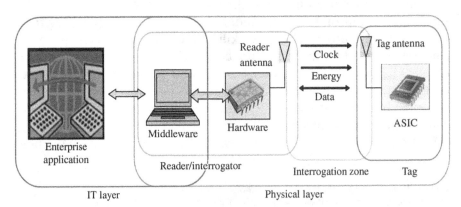

Figure 1.1   Architecture of conventional radio frequency identification system.

(ASIC) or the microchip of the tag. The removal of chip from the tag will lower the cost of tag to a great extent. This can be an excellent alternative for traditional barcodes, which suffer from several issues such as the following: (a) each barcode is individually read, (b) needs human intervention, (c) has less data handling capability, (d) soiled barcodes cannot be read, and (e) barcodes need line-of-sight operation. Despite these limitations, the low-cost benefit of the optical barcode makes it very attractive as it is printed almost without any extra cost. Therefore, there is a pressing need to remove the ASIC from the RFID tag to make it competitive in mass deployment. After removing the ASIC from the RFID tag, the tag can be printed on paper or polymer, and the cost will be less than a cent for each tag [4]. The IDTechEx research report [2] advocates that 60% of the total tag market will be occupied by the chipless tag if the tag can be made less than a cent. As most of the tasks for RFID tag are performed in the ASIC, it's not a trivial task to remove it from the tag. It needs tremendous investigation and investment in designing low-cost but robust passive microwave circuits and antennas using conductive ink on low-cost substrates. Additionally to these, obtaining high-fidelity response from low-cost lossy materials is very difficult [4]. In the interrogation and decoding aspects of the RFID system is the development of the RFID reader, which is capable to read the chipless RFID tag. Conventional methods of reading RFID tags are not implementable in reading chipless RFID tags. Therefore, dedicated chipless RFID tag readers need to be implemented [5]. This is the first book in this discipline that presents detailed aspects, challenges, and solutions for advanced signal processing for chipless RFID readers for detection, tracking, and anticollision.

The market of chipless RFID is emerging slowly, and the demand is increasing day by day. As forecasted by IDTechEx, the market volume of chipless RFID was less than $5 million in 2009. However, this market will grow to approximately $4 billion in 2019 [6, 7]. In contrast to 4–5 billion optical barcodes that are printed yearly, approximately 700 billion chipless tags will be sold in 2019. Therefore, a significant interest is growing in researchers for the development and implementation of low-cost chipless RFID systems. This book is presenting the advanced signal processing methods that are being used in chipless RFID system for detection, identification, and tracking, and collision avoidance.

The development of chipless RFID systems has already come a long way. Compared to early days, it has already in its second-generation development phase with more data capacity, reliability, and compliances of some existing standards. RF-SAW tags got new standards, can be made smaller with higher data capacity, and currently are being sold in millions. Approximately 30 companies have been developing TFTC. TFTC targets HF (13.65 MHz) band (60% of existing RFID market) and has the read–write capability [7].

In generation I, only a few chipless RFID tags, which were in the inception stage, were reported in the open literature. They include a capacitive gap coupled dipole array [8], a reactively loaded transmission line [9], a ladder network [10], and finally a piano and a Hilbert curve fractal resonators [11]. These tags were in prototype stage, and no further development in commercial grade was reported so far.

It is obvious that chipless RFID is a potential option for replacing barcodes and hence realizing the fact big industry players such as IBM, Xerox, Toshiba, Microsoft, HP, and new players such as Kavio and Inksure have been investing tremendously in the development of low-cost chipped and chipless RFID. Figure 1.2 shows the motivational factors in developing chipless RFID tags and reader systems. The data shown in the figure is approximated from two sources [6, 7]. Currently, the conventional chipped tags cost more than 10¢ if purchased in large

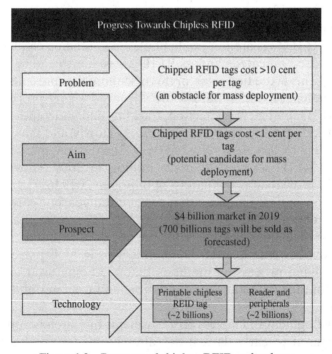

Figure 1.2 Prospect of chipless RFID technology.

quantities. This high tag price hinders mass deployment of RFID in low-cost item-level tagging. The goal is to develop sub-cent chipless tags that will augment the low-cost item-level tagging. The technological advancements in both the chipless tags and their readers and peripherals will create approximately $4 bn market in 2019 [6, 7].

According to the prediction of www.MarketsandMarkets.com (accessed on June 9, 2012) [6], the revenue generated in global chipless RFID market is expected to reach $3925 million in 2016 from $1087 million in 2011, at an estimated combined annual growth rate of 29.3% from 2011 to 2016. The targeted market sectors for the chipless RFID include retail, supply chain management, access cards, airline luggage tagging, aged care and general healthcare, public transit, and library database management system. The author's group has been developing chipless RFID tag technologies targeting many of these sectors since 2004.

To the best of the author's knowledge, there is only one book so far by the same author group regarding the chipless RFID reader development [12]. The published book mainly focuses on the hardware development and implementation for chipless RFID tag reader. However, the background signal processing and identification have not been discussed in detail. This book focuses on the signal postprocessing for tag identification, tracking, noise mitigation, and multi-tag identification aspects. The author group and their chipless RFID research team has been working on the chipless RFID tag readers since 2004 [13]. Significant strides have been achieved to tag not only the polymer banknotes but also many low-cost items such as books, postage stamps, secured documents, bus tickets, and hanging cloth tags. The technology relies on encoding spectral signatures and decoding the amplitude and phase of the spectral signature [14]. The other methods are phase encoding of backscattered spectral signals [15] and time-domain delay lines [16]. So far, as many as more than twenty varieties of chipless RFID tags and five generations of readers are designed by this team. The proof of concept technology is being transferred to the banknote polymer and paper for low-cost item tagging. These tags have potential to coexist or replace trillions of optical barcodes printed each year. To this end, it is imperative to invest on low-cost conductive ink, high-resolution printing process, and characterization of laminates on which the tag will be printed. The design of a spectral signature-based tag needs to push in higher frequency bands to accommodate and increase the number of bits in the chipless tag to compete with the optical barcode. In this space, the reader design needs to accommodate large reading distance and high-speed reading, multiple tag reading in close proximity, error correction coding, and anticollision protocols. Also, wide acceptance of RFID technology by consumers and businesses requires robust privacy and security protection [16, 17]. The book aims to address all these issues mentioned earlier to make the chipless RFID system a viable commercial product for mass deployment.

Figure 1.3 shows the salient features of a chipless RFID tag, and Figure 1.4 shows the burdens of a chipless RFID tag reader to meet the market demands. It is a highly challenging and interesting task to design a dedicated chipless RFID tag reader with all the requirements fulfilled as well as cost-effective. Figure 1.4 shows the chipless RFID system, which needs to address a whole spectrum of technical and regulatory requirements such as the number of data bits to be read and processed, operating frequencies, radiated (transmitted) power levels, and hence reading distance, mode of readings (time, frequency, or hybrid domain), compatibility with existing technological framework, simultaneous multiple tag reading, and resulting anticollision and security issues. All of these considerations will impact the development and commercialization of the new technology. IDTechEx [7] reports on the chipless RFID tag development by commercial entities and highlights the synergies to address all these issues to make the chipless RFID a commercially viable and competitive technology. The objective of the book is to address many significant issues and provide technical solutions of chipless RFID readers.

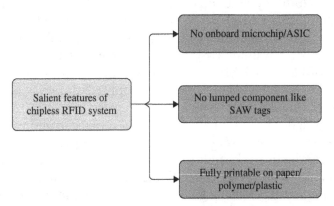

Figure 1.3 Salient features of chipless RFID tag (© 2010 Editor © Karmakar, N. C. Published 2010 by John Wiley & Sons, Ltd.).

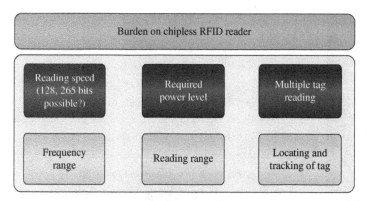

Figure 1.4 Design considerations for chipless RFID tag reader.

1.2 CHIPLESS RFID TAG READER

As advocated by IDTechEx [7], the chipless RFID tag readers and data processing software will cost similar to their chipped counterparts. The market segment of hardware and related middleware is more than the cost of tags used. Therefore, there are huge commercial potentials to invest in readers and related software development. However, there are no such resources on chipless RFID reader systems in the open litera-ture as yet. This is the first initiative by the author to introduce the potential field in a combined and comprehensive body of literature. The book covers the following topics in the field in five sections:

1. Introduction to chipless RFID
2. Chipless RFID tag detection techniques
3. Noise mitigation for improving detection accuracy
4. Multi-tag identification and collision avoidance
5. Tag localization and tracking

Figure 1.5 shows the organization of the chapters in this book. The topics covered in each chapter are highlighted below:

Section 1: Detection and Denoising

Some fundamental system-level issues of the chipless RFID tag reader systems and their detection methods are presented in Chapters 2–5. In an efficient system, raw data obtained from the RF transceiver of the

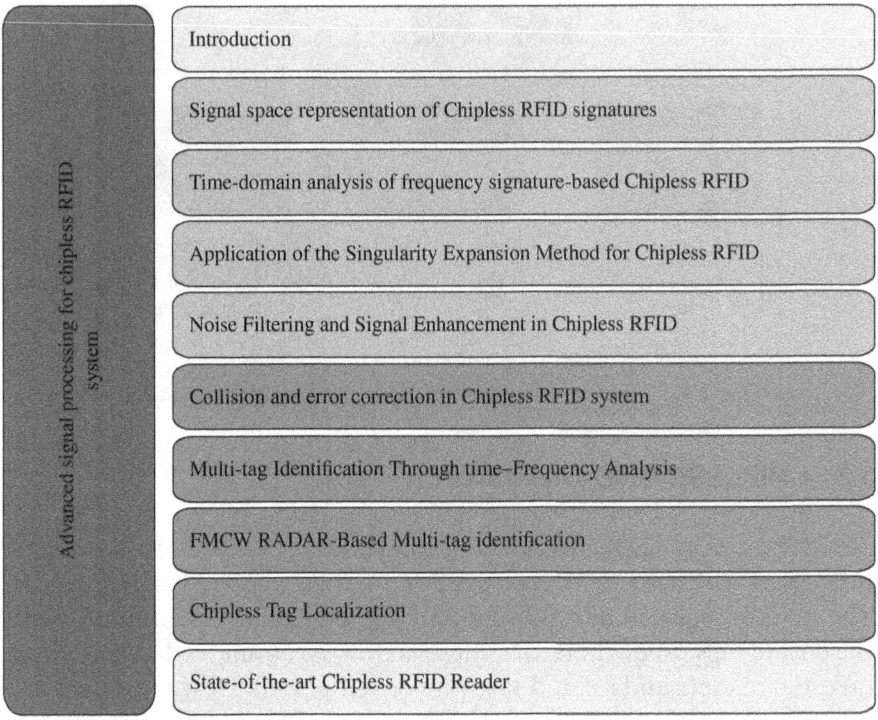

Figure 1.5   Organization of sections and chapters of chipless RFID systems.

chipless reader needs to be processed and denoised. In this aspect, a few new techniques are reported in this section. They are (i) signal space representation of chipless RFID signatures, (ii) detection of frequency signature-based chipless RFID using UWB impulse radio interrogation, (iii) singularity expansion method (SEM) for data extraction from chipless RFID, and finally (iv) noise reduction and filtering techniques used for the chipless RFID. These methods improve the efficacy and throughput of various types of reading processes. The following are the detailed descriptions of the chapters.

Chapter 2: Signal Space Representation of Chipless RFID Signatures

In this chapter, the decoding of information contained in a chipless RFID tag signature is realized using signal space representation technique. This method is commonly used in conventional digital communication systems. Therefore, this is a new approach to chipless RFID detection. A different perspective on the detection of chipless RFID is presented where a mathematical model based on signal space representation is used. Here, the chipless RFID tag's signatures are visualized as signal points in a signal space (Euclidian space), which enables the detection of data bits contained in these signatures through conventional methods used in digital communications. It is shown that the proposed method has better performance compared to a threshold-based approach to detection.

Chapter 3: Time-Domain Analysis of Frequency Signature-Based Chipless RFID

This chapter presents the use of UWB impulse radio interrogation to remotely estimate the frequency signature of chipless RFID tags that are operating using the backscatter principle. Two types of frequency signature-based chipless RFID tags are investigated: (i) a multiresonator-loaded chipless RFID tag [14] and (ii) a multipatch-based chipless RFID tag [15]. Here, the received signal from a frequency signature-based chipless RFID tag is captured in the time domain. The spectral contents of the tag's returned echo signal are analyzed to identify the key components that make up the information-carrying signal. It is shown that the information-carrying portion of the signal is contained in the *antenna mode* of the backscattered signal, and the *structural mode* of the backscatter holds no information about the resonant features of the tag [18, 19]. The performance of the method is investigated under different tag positions. The theory of operation is validated using simulation, semianalytical methods, and experimental results.

Chapter 4: Singularity Expansion Method for Data Extraction for Chipless RFID

This chapter details the use of the SEM for extracting information from the chipless RFID tag. The theory of the SEM is reviewed, and its application to chipless RFID is explained. The SEM technique is used to characterize the response of an object that is subjected to a burst of high-energy electromagnetic (EM) radiation. Here, the transient response of an object that is excited by an impulse of EM energy is characterized using a set of poles and residues. Several works reported in literature that are based on the application of SEM for chipless RFID research are discussed.

Chapter 5: Denoising and Filtering Techniques for Chipless RFID

Backscatter received from chipless RFID tags are very weak and are affected by many detriments that make detection very challenging. These detriments include the additive thermal noise introduced by reader electronics, interfering echoes caused by clutter in the environment, and transient responses of receive and transmit antennas. The detection of weak signals from chipless RFID tags amidst these factors requires techniques to enhance the ratio of signal level to the interference and noise level. This chapter details some of the techniques reported in literature for detecting signals from chipless RFID that are contaminated by noise. Specifically, the use of wavelet transforms and prolate spheroidal wave functions for noise filtering is discussed.

Section 2: Multiple Access and Localization in Chipless RFID

The main motto of developing chipless RFID technology is to reduce the cost of the tag in sub-cent level and to facilitate mass deployment of the RFID technology for low-cost item-level tagging. Therefore, multiple access with collision avoidance techniques for proximity tags and subsequent signal integrity is a significant research. Unfortunately, the protocols dedicated for these purposes in the conventional chipped tag cannot be applied in chipless RFID tag scenarios. The reason behind this is that the chipless RFID tags do not contain an ASIC microchip capable of signal processing. Since the chipless tag is a fully printable microwave passive design and void of any intelligence, the processing for multiple access and signal integrity is done only in the chipless RFID tag reader. So far, little has been achieved in this field. In this book, three chapters are dedicated to address the issue and full up the gap in the open literature. A comprehensive background and possible methods of multiple access and system integrity are presented. Chapters 6, 7, and 8 present linear block coding, time–frequency analysis, and finally frequency-modulated continuous wave (FMCW) radar for various chipless RFID tags. The detailed summaries of the chapters are presented below:

Chapter 6: Collision and Error Correction Protocols in Chipless RFID

Collision is an inherent problem in wireless communications. This chapter reviews the collision problem in both chipped and chipless RFID systems and summarizes the prevailing anticollision algorithms to address the problem. Due to the uniqueness of chipless RFID system, the available collision avoidance methods are not applicable to the chipless RFID reader. Therefore, a collision avoidance method based on linear block coding for frequency-domain chipless RFID tags is discussed and some preliminary simulation results as the proof of concept are presented in the chapter.

Chapter 7: Multi-Tag Identification through Time–Frequency Analysis

Time–frequency analysis is an excellent tool for analyzing time-varying signals. It is widely used in radar signal processing. This chapter describes the potential application of time–frequency analysis, especially fractional Fourier transform (FrFT) in the chipless RFID system [20]. Instead of analyzing the collided response signal in the frequency domain, it is converted to time–frequency plane. FrFT compresses individual response signals to different regions in fractional plane. Afterward, windowing is used to separate them. The tag ID is determined from the spectrum of the separated signal. The validation of the algorithm has been carried out through simulation for different tag combinations.

Chapter 8: FMCW-Radar-Based Multi-Tag Identification

This chapter describes the application of FMCW-radar technique for localization, collision detection, and multiple access in the chipless RFID system. For localization, multiple antennas are placed around the tag to calculate the round-trip time-of-flight (RTOF) of response signal. Based on RTOF, trilateration localization technique is employed to localize the tag in two-dimensional plane. The signals from multiple tags are downconverted to intermediate frequency (IF) signal, and the spectrum is analyzed for collision detection and number of collided tag estimation. Afterward, each beat frequency signal corresponding to a tag is filtered out and analyzed for resonance information.

Chapter 9: Chipless Tag Localization

This chapter presents a localization method for chipless RFID tag. A short-duration ultra-wideband impulse radio (UWB-IR) signal interrogates the tags, and multiple receivers in the interrogation zone capture the backscattered signal from the tags. The received signals from the chipless tags are analyzed for the structural mode radar cross section

(RCS) to determine the relative ranges. Using the range information, Linear Least Square (LLS) method is employed for accurate localization of tagged items. The accuracy of localization method is analyzed by moving the chipless tag within a fixed interrogation zone. The analysis and results create a strong foundation for chipless RFID tags to be used in tracking and localization [21].

1.3 CONCLUSION

RFID is an emerging technology for automatic identification, tracking, and tracing of goods, animals, and personnel. In recent decades, the exponential growth of RFID market signifies its potentials in numerous applications. The advantageous features and operational flexibility of RFID have attracted many innovative application areas. Therefore, there is a need for tremendous development and open literature on RFID to report new results. However, the bottleneck of mass deployment of RFID technology is the cost of the tag and reading techniques and processes of the chipless tag. This is the first book in the field that covers comprehensively many significant aspects of chipless RFID reader architecture and signal processing.

This book project is an initiative to publish most recent results of research and development on the chipless RFID tag reader system. It aims to serve the needs for a broad spectrum of readers. The book has proposed a few novel detection techniques dedicated to radar array-based chipless RFID tags [14, 15]. Firstly, the signal space representation-based chipless RFID tag detection is used for the frequency signature-based chipless tags. Secondly, a UWB impulse radio detection technique is used to interrogate a frequency signature-based chipless tag. Thirdly, a SEM is used to separate poles and residues of the tag responses, and finally, various filtering techniques such as wavelet transformation and prolate spheroidal wave function are used for noise filtering. All these detection, denoising, and filtering techniques improve the efficacy of the reader.

The book has also proposed a few state-of-the-art multi-access and signal integrity protocols to improve the efficacy of the system in multiple tag reading scenarios. Comprehensive studies of anticollision protocols for the chipless RFID systems and various revolutionary techniques to improve the signal integrity have also been presented. It has also identified future challenges in the sphere. It also discusses the location finding of chipless tag and hence tracing and tracking of tagged items.

Finally, an industry approach to the integration of various systems of the chipless RFID reader technology—integration of physical layers, middleware, and enterprise software—is the main feature of the book. Overall, the book has become a one-stop shop for a broad spectrum of readers who have interests in the emerging chipless RFID and sensor technologies.

REFERENCES

1. IDTechEx, *Piezoelectric Energy Harvesting 2012–2022: Forecasts, Technologies, Players*, August 2, 2012. http://www.reportsnreports.com/reports/163578-piezoelectric-energy- harvesting-2012-2022-forecasts-technologies-players.html; http://www.reportlinker.com/p0944823/Piezoelectric-Energy-Harvesting-2012-2022-Forecasts-Technologies-Players.html (accessed July 7, 2015).

2. IDTechEx, *RFID Forecasts, Players and Opportunities 2009–2019, Executive Summary*, April, 2009. http://www.reportlinker.com/p0149567/Reportlinker-Adds-RFID-Forecasts-Players-and-Opportunities-2009-2019.html (accessed July 7, 2015).

3. K. Finkenzeller, *RFID Handbook: Fundamentals and Applications in Contactless Smart Cards, Radio Frequency Identification and Near-Field Communication*, 2nd Ed., 2003, John Wiley & Sons, Inc., New York.

4. R. E. Azim, N. C. Karmakar, S. M. Roy, R. Yerramilli, and G. Swiegers, "Printed Chipless RFID Tags for Flexible Low-Cost Substrates," in: *Chipless and Conventional Radio Frequency Identification: Systems for Ubiquitous Tagging*, Editor: N. C. Karmakar, IGI Global, Hoboken, NJ, 2012.

5. S. Parardovic and N. C. Karmakar, "RFID Readers—Review and Design," in: *Handbook of Smart Antennas for RFID Systems*, Wiley Microwave and Optical Engineering Series, John Wiley & Sons, Inc., Hoboken, NJ, 2010, pp. 85–122, 2010.

6. marketsandmarkets.com, *Chipless RFID Market (2011–2016)—Global Forecasts by Applications (Retail, Supply Chain, Aviation, Healthcare, Smart Card, Public Transit & Others)*, SE 1714, November 2011. http://www.marketsandmarkets.com/Market-Reports/chipless-rfid-market-501.html (accessed June 9, 2012).

7. IDTechEx, *Printed and Chipless RFID Forecasts, Technologies & Players* 2009–2029, 2009. http://media2.idtechex.com/pdfs/en/R9034K8915.pdf (accessed July 7, 2015).

8. I. Jalaly and I. D. Robertson, "Capacitively Tuned Microstrip Resonators for RFID Barcodes," *35th 2005 European Microwave Conference, EUMC, Vol. 2*, Paris, October 4–6, (pp. 1161–1164), 2005.

9. L. Zhan, S. Rodriguez, H. Tenhunen, and L. R. Zheng, "An Innovative Fully Printable RFID Technology Based on High Speed Time-Domain Reflections," *Conference on High Density Microsystem Design and Packaging and Component Failure Analysis, 2006. HDP'06*, Shanghai, June 27–30, (pp. 188–198) 2006.

10. S. Mukherjee, "Chipless Radio Frequency Identification by Remote Measurement of Complex Impedance," *2007 European Conference on Wireless Technologies*, Munich, October 8–10, (pp. 249–252), 2007.

11. J. McVay, A. Hoorfar, and N. Engheta, "Theory and experiments on Peano and Hilbert curve RFID tags," *Proceedings of the Society for Photo-Instrumentation Engineers, SPIE, vol. 6248, no. 1*, San Diego, August, 2006, (pp. 624808-1–624808-10), 2006.

12. N. C. Karmakar, R. V. Koswatta, P. Kalansuriya, and R. Azim, *Chipless RFID Reader Architecture*, Artech House, Boston, 2013.

13. Karmakar, N. C. (2010). *Handbook of Smart Antennas for RFID Systems*. Wiley Microwave and Optical Engineering Series, John Wiley & Sons, Inc., Hoboken, NJ

14. S. Preradovic, I. Balbin, N. C. Karmakar, and G. F. Swiegers, "Multiresonator-based chipless RFID system for low-cost item tracking," *IEEE Transactions on Microwave Theory and Techniques*, vol. 57, no. 5, Part 2, 2009, pp. 1411–1419.

15. I. Balbin and N. C. Karmakar, "Multi-Antenna Backscattered Chipless RFID Design," in: *Handbook of Smart Antennas for RFID Systems*, Wiley Microwave and Optical Engineering Series, John Wiley & Sons, Inc., Hoboken, NJ, pp. 415–444.

16. R. Woolf, *Development of A 2.45 GHz Chipless Transponder for RFID Application*, Final Year Project Thesis, ECSE, Monash University, 2009.

17. J. Kamruzzaman, A. K. M. Azad, N. C. Karmakar, G. C. Karmakar, and B. Srinivasan, "Security and Privacy in RFID System," in: *Advanced RFID Systems, Security, and Applications*, IGI Global, Hoboken, NJ, 2012.

18. P. Kalansuriya and N. C. Karmakar, "UWB-IR based detection for frequency-spectra based chipless RFID," *2012 International Microwave Symposium*, Montreal, Canada, June 17–22, 2012 (acceptance date June 2, 2012).

19. P. Kalansuriya and N. C. Karmakar, "Time domain analysis of a backscattering frequency signature based chipless RFID tag," *2011 Asia Pacific Microwave Conference Proceedings*, Melbourne, Australia, December 5–8, 2011.

20. R. Azim and N. Karmakar, "A Collision Avoidance Methodology for Chipless RFID Tags," *2011 Asia Pacific Microwave Conference Proceedings*, Melbourne, Australia, December 5–8, 2011.

21. R. Anee and N. C. Karmakar, "Chipless RFID Tag Localization," *IEEE Transactions on Microwave Theory and Techniques*, vol. 61, no. 11, pp. 4008–4017, 2013.

CHAPTER 2

SIGNAL SPACE REPRESENTATION OF CHIPLESS RFID SIGNATURES

2.1 WIRELESS COMMUNICATION SYSTEMS AND CHIPLESS RFID SYSTEMS

This section presents a contrast between the conventional digital wireless communication used in Bluetooth, wireless local area networks (WLAN), or chipped RFID tag systems and the nonstandard passive backscatter-based wireless communication utilized in chipless RFID systems.

2.1.1 The Conventional Digital Wireless Communication System

A conventional digital wireless communication system, shown in Figure 2.1, comprises of a transmitter, a receiver, and a wireless medium (air) between them. The transmitter node sends information through the wireless medium to the receiver node. For two nodes in a WLAN, the role of transmitter and receiver will interchange depending on which node is requesting information. This information, which is in the form of digital data bits, is modulated on to a radio frequency carrier wave using a modulation scheme such as quadrature amplitude modulation (QAM)

Chipless Radio Frequency Identification Reader Signal Processing, First Edition.
Nemai Chandra Karmakar, Prasanna Kalansuriya, Rubayet E. Azim and Randika Koswatta.
© 2016 John Wiley & Sons, Inc. Published 2016 by John Wiley & Sons, Inc.

Figure 2.1   A conventional wireless communication system consisting of a transmitter, a wireless channel, and a receiver.

[1] and is transmitted through the wireless channel. The wireless channel between the transmitter and receiver serves as an obstacle to reliable and high-speed data communication. It introduces path loss, multipath fading, and shadowing [1] to the signal. Due to these factors, the received signal at the receiver experiences signal degradation and loss of integrity. In addition to these detriments, the receiver electronics and RF front end will introduce thermal noise, which will further deteriorate the received signal. This noise is often modeled as additive white Gaussian noise (AWGN) in literature [1]. However, amidst all these complicated factors adversely affecting the received signals, it is still possible to achieve very high-quality, high data rate wireless digital transmission using modern wireless transmission standards such as the IEEE 802.11 and the Long-Term Evolution (LTE) technology [2]. The key to such wireless fidelity lies in the fact that the receiver resorts to estimating the wireless channel before any useful data transmission takes place and also the use of state-of-the-art techniques such as orthogonal frequency-division multiplexing (OFDM) [1]. With channel estimation, it is possible to combat the detrimental effects of the wireless channel and reliably detect the information contained in the received signal. Channel estimation is performed by sending data symbols that are known beforehand to both the transmitter and receiver and by observing how the wireless channel distorts these symbols. These symbols are often called pilot symbols. Using the responses obtained for the transmitted *pilot symbols*, the receiver estimates the effect of the wireless channel and cancels its effect for all the successive data transmissions from the transmitter. This process of neutralization of the adverse effects of the channel is also called *equalization*.

2.1.2   Chipped RFID System

The communication between an RFID reader and a chipped RFID tag is also performed using conventional digital communication. Even the most primitive of the chipped RFID tags, the passive backscatter-based tags, conform to these principles. In a passive chipped RFID tag, the RFID reader basically supplies the radio frequency carrier wave that both energizes the RFID chip in the tag and also serves as the modulating medium onto which the chip actively modulates its digital electronic product code. Here, the reader serves as the receiver, and the tag serves as the transmitter in the model shown in Figure 2.1. Simple modulation schemes such as frequency-shift keying (FSK), amplitude-shift keying (ASK), or phase-shift keying (PSK) are used since the electronic circuitry of a passive RFID tag is not capable of handling advanced modulation techniques.

2.1.3 Chipless RFID System

In contrast to a conventional communication system, in a chipless RFID system, the entity that holds the information, which needs to be conveyed to a reader, does not possess any intelligence or capability to transmit it. Therefore, in order to read the chipless RFID tag, a different and nonstandard wireless communication protocol is used [3]. Here, both the transmitter and the receiver are colocated in the RFID reader as shown in Figure 2.2. The transmitter interrogates the chipless RFID tag with an interrogation signal, and the colocated receiver of the RFID reader listens to the backscatter response coming back from the tag. The chipless RFID tag essentially transforms the interrogation signal to carry back its information to the reader. However, compared to the chipped passive RFID system, these transformations are neither performed actively by an electronic chip nor do they rely on digital modulation techniques. The transformations are completely passive and occur due to the nature of design of the chipless RFID tag.

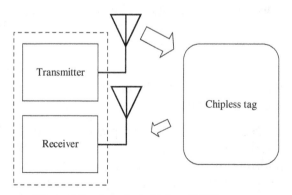

Figure 2.2 A typical chipless RFID system.

For example, a chipless tag that is based on planar microwave filters [4, 5] transforms the interrogation signal in the frequency domain causing some frequencies to sharply attenuate for representing its data, whereas a chipless tag based on microwave meander transmission lines [6–8] introduces time lags or multipath to the interrogation signal to represent its data. At any rate, the process of reading the information contained in a chipless RFID tag involves in sending a known signal and observing how the chipless tag transforms it. This reading process is essentially equivalent to the channel estimation process described for a conventional communication system earlier. However, the main difference is that in a conventional communication system, the channel estimation is performed in order to remove the detriments of the wireless channel and reliably transmit data across it, whereas in a chipless RFID system, the estimated channel is essentially the data that needs to be read, and the transmitted signals serve only as the pilot symbols used for the channel estimation [9].

2.2 THE GEOMETRIC REPRESENTATION OF SIGNALS IN A SIGNAL SPACE

This section introduces the concept of signal space representation [9, 10] where continuous time signals are geometrically represented as vectors (or signal points) in a vector space (or a signal space). These concepts are readily used in digital and wireless communications in modeling and abstracting the complicated mechanisms involved with the transmission and reception of information through a communication channel. Here, we aim to provide some background theory on the use of these concepts in the context of digital communication. This context will be useful in understanding and contrasting how signal space representation is used for the detection of information in a chipless RFID tag later on.

2.2.1 Representing Transmit Signals Using Orthonormal Basis Functions

At the transmitter, it is desirable to represent the signal being transmitted using a set of orthonormal basis functions. Let $\phi_i(t), i = 1,...,N$ be a set of orthonormal functions, which are time limited and exist from $0 < t < T$. These set of functions are called orthonormal due to two special properties they possess, which are defined as follows [11]:

$$\int_0^T \phi_i(t)\phi_j(t)\,dt = \begin{cases} 1; & i=j \\ 0; & i \neq j \end{cases} \tag{2.1}$$

Equation 2.1 can also be written using the inner product notation generally used in describing vector spaces [12], that is, $\langle \phi_i(t), \phi_j(t) \rangle = 0$ and $\langle \phi_i(t), \phi_i(t) \rangle = \|\phi_i(t)\|^2 = 1$. These equations simply state that all these functions are orthogonal to each other where they can occupy the same time period simultaneously with the ability of being completely decoupled of each other so that each function can be recovered perfectly without being affected by the others; for example, you can consider unit vectors along the three dimensions length, height, and width to be orthonormal, where each vector would exist independently in the same space without affecting the other. When the signal being transmitted is written in terms of a set of orthonormal functions, it is possible to send information simultaneously using each of the orthogonal dimensions presented by each of the orthonormal functions, which results in higher spectral efficiency. Also, it would help in combating against AWGN.

Let $x_k(t), 0 < t < T$ be the kth information-carrying signal being transmitted by the transmitter where each $x_k(t)$ can be expressed as a weighted sum of $\phi_i(t)$ as follows:

$$x_k(t) = \sum_{i=1}^N a_{i,k}\phi_i(t) \tag{2.2}$$

where $a_{i,k}$ is the scalar weight that defines the amplitude of the orthonormal function, $\phi_i(t)$, contributing to produce $x_k(t)$. When the time notation, "t," in the signal is dropped, $x_k(t)$ can be represented as a vector $\mathbf{x}_k$, where each information-carrying $\mathbf{x}_k$ is expressed using a unique linear combination of the orthonormal set of vectors $\varphi_i, i = 1, \ldots, N$:

$$\mathbf{x}_k = \sum_{i=1}^N a_{i,k}\varphi_i \tag{2.3}$$

Therefore, now, we can represent each of the continuous time information-carrying signals $x_k(t)$ as vectors or signal points geometrically in a Euclidian vector space as shown in Figure 2.3.

Figure 2.3a shows a two-dimensional signal space ($N=2$) defined using two orthonormal vectors φ_1 and φ_2. Three information-carrying signals, $x_k(t), k = 1, 2, 3$, are represented as vectors in the signal space where $x_1(t) = a_{1,1}\phi_1(t) + a_{2,1}\phi_2(t)$, $x_2(t) = a_{1,2}\phi_1(t) + a_{2,2}\phi_2(t)$, and

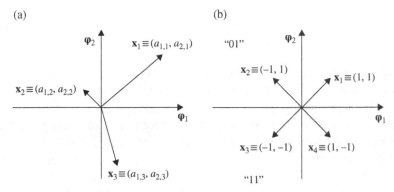

Figure 2.3 Signal space representation of signals. (a) Three information-carrying signals, $x_1(t) = a_{1,1}\phi_1(t) + a_{2,1}\phi_2(t)$, $x_2(t) = a_{1,2}\phi_1(t) + a_{2,2}\phi_2(t)$, and $x_3(t) = a_{1,3}\phi_1(t) + a_{2,3}\phi_2(t)$ represented in a two-dimensional vector space as vectors $\mathbf{x}_1$, $\mathbf{x}_2$, and $\mathbf{x}_3$. (b) The signal constellation of 4-QAM consisting of four vectors.

$x_3(t) = a_{1,3}\phi_1(t) + a_{2,3}\phi_2(t)$. By changing these weights, the information carried by each $x_k(t)$ can be controlled. These weights can have discrete values in order to encode digital information into $x_k(t)$. Using a pre-defined set of coefficients $a_{i,k}$, 2^b distinct signals, $x_k(t)$, $k = 1,\ldots,2^b$, can be constructed. The transmitter can then encode "b" data bits in each of these signals. These sets of signals $x_k(t)$, $k = 1,\ldots,2^b$ can be considered as a signal constellation that can be used to represent "b" data bits. For example, Figure 2.3b shows the 4-QAM constellation that consists of four signals where each signal represents two data bits. In the process of transmitting a stream of digital data bits, the transmitter breaks it up into groups of "b" bits and uses a sequence of appropriate signals, $x_k(t)$, in order to represent the stream of data.

2.2.2 Receiving Signals and Decoding Information

Assuming that the transmitted signals are communicated through an AWGN channel where the only detriment affecting the communication is noise, the received signal can be written as

$$y(t) = x_k(t) + n(t) \tag{2.4}$$

where $n(t)$ is the additive noise affecting the signal. When expressed using the vector notation as in Equation 2.3, the received vector can be written as

$$\mathbf{y} = \sum_{i=1}^{N} a_{i,k}\phi_i + \mathbf{n} \tag{2.5}$$

The goal of the receiver is to estimate with least amount of error the scalar coefficients $a_{i,k}$ that holds the information contained in $x_k(t)$. The estimation is performed by taking the inner product between the received vector (2.5) and each of the orthonormal basis vectors, φ_j, $j = 1,...,N$:

$$\langle \mathbf{y}, \varphi_j \rangle = \sum_{i=1}^{N} a_{i,k} \langle \varphi_i, \varphi_j \rangle + \langle \mathbf{n}, \varphi_j \rangle \qquad (2.6)$$

All the cross terms in the summation of Equation 2.6 reduce to zero because $\langle \varphi_i, \varphi_j \rangle = 0$ for $i \neq j$ where only $a_{j,k}$ and the noise term $\langle \mathbf{n}, \varphi_j \rangle$ remains, since for $i = j$, $\langle \varphi_i, \varphi_j \rangle = \| \varphi_j \|^2 = 1$. Therefore, Equation 2.6 reduces to

$$\langle \mathbf{y}, \varphi_j \rangle = a_{j,k} + \langle \mathbf{n}, \varphi_j \rangle, \quad j = 1,...,N \qquad (2.7)$$

The inner product between $\mathbf{y}$ and φ_j, $\langle \mathbf{y}, \varphi_j \rangle$, is the projection of the received vector $\mathbf{y}$ onto the φ_j axis of the signal space. In other words, $\langle \mathbf{y}, \varphi_j \rangle$ is the jth element or coordinate, y_j, of the received vector $\mathbf{y}$, which represents the amount of the unit vector φ_j in $\mathbf{y}$; that is, $\mathbf{y} \equiv (y_1, y_2,...,y_j,...,y_N)$ or $\mathbf{y} = \sum_{j=1}^{N} y_j \varphi_j$. The term $\langle \mathbf{n}, \varphi_j \rangle$ is the amount of noise projected onto the φ_j dimension of the signal space. Figure 2.4 illustrates the signal processing structure at the receiver used for estimating $x_k(t)$ from $y(t)$.

When the statistics of the noise vector is known a priori, it is possible to estimate the transmitted data signal using the received vector with minimal error. Figure 2.5 illustrates the received vector in the signal space together with the signal constellation used for data transmission. When $\mathbf{x}_k$ is transmitted due to the noise introduced by the channel, the received vector $\mathbf{y}$ has displaced from the intended location in the constellation. Since the receiver does not know what signal was transmitted, the estimation block in Figure 2.4 should be able to estimate the transmitted signal using only $\mathbf{y}$. From the figure, intuitively, it is highly likely that the received vector $\mathbf{y}$ is due to the transmission of $\mathbf{x}_k$ since it is closer to the constellation point corresponding to $\mathbf{x}_k$. The derivation of the exact optimal estimation procedure requires knowledge of the statistics of the noise process affecting the transmitted signal and the decision boundaries that are used in demarcating the regions corresponding to each of the constellation points in the signal constellation [11].

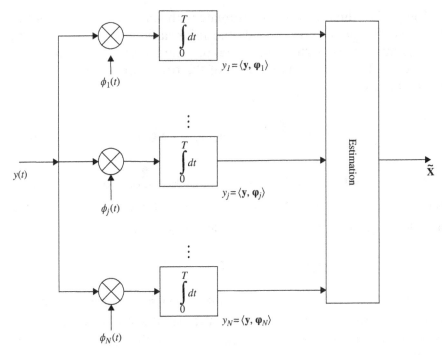

Figure 2.4   The signal processing structure at the receiver [11].

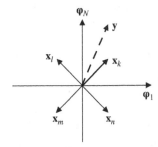

Figure 2.5   The received signal vector plotted in the signal space together with the signal constellation.

2.3   NOVEL MODEL FOR THE REPRESENTATION OF CHIPLESS RFID SIGNATURES

This section presents the formulation of a model for representing chipless RFID signatures in a signal space [9, 10]. This model enables the detection of information bits embedded in the RFID signatures. For the purpose of demonstrating the concept, a frequency signature-based chipless RFID system is considered. Figure 2.6 shows the system model of the chipless RFID system. The system uses a retransmission-based chipless RFID tag [5].

The retransmission-based chipless RFID tag comprises of three components: a receiving antenna to receive the interrogation signal sent from the reader, a passive microwave filter capable of transforming the received interrogation signal, and a transmitting antenna to retransmit the transformed signal back toward the reader. The operation is quite straightforward; the RFID reader transmits the interrogation signal, $X(f)$, which travels through the forward wireless channel $W_F(f)$ and is picked up by the receiving antenna of the tag. The received signal then travels through the passive microwave filter of the tag and is transformed to carry the information stored in the tag. This transformation is characterized by the transfer function $H_k(f)$ of the microwave filter. Abrupt variations in the amplitude or phase of $H_k(f)$ are used in representing information bits. Hence, different transfer functions $H_k(f)$, $k = 1,\dots,\ 2^b$ are used to represent different data, where 2^b different $H_k(f)$s are required to construct a "b" bit chipless RFID system. This transformed interrogation signal is retransmitted via the transmitting antenna of the chipless tag through the reverse wireless channel $W_R(f)$. The signal that reaches the receiving antenna of the reader is very weak and will be adversely affected by thermal noise $N(f)$. We assume that there exists a direct unobstructed line-of-sight path between the reader and the tag. Therefore, the forward and reverse wireless channels only introduce a time delay and attenuation (path loss) to the signals traveling through them. Apart from the adverse effects of the thermal noise, there exists another deterrent to the accurate detection of

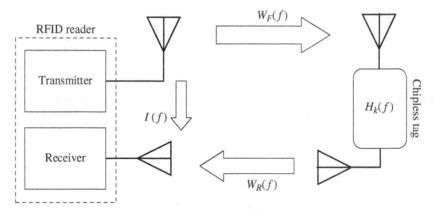

Figure 2.6 The system model of a retransmission-based chipless RFID system.

information contained in the chipless tag, which is the coupling between the transmit and receive antennas of the RFID reader and the clutter produced by the surrounding environment. By having the reader antennas in a cross polarized configuration, this coupling component can be minimized. However, when the distance to the tag is increased, the signal level received at the reader becomes weaker and comparable in magnitude with the coupling component. This causes ambiguity and detection problems at the reader. Taking all these factors into consideration, the total received signal, $Y(f)$, at the reader receiving antenna can be expressed as

$$Y(f) = X(f)W_F(f)H_k(f)W_R(f) + I(f)X(f) + N(f) \quad (2.8)$$

where $I(f)$ is the transfer function that describes the behavior of the interference caused by the coupling and the clutter. As explained in the previous sections, the information stored in the chipless RFID tag is contained in the transfer function $H_k(f)$ where the transmitted signal $X(f)$ only serves as a pilot signal used for estimating $H_k(f)$. Using calibration measurements such as empty measurements without the tag or using a calibration tag measurement [13], the effect of the interfering component, $I(f)X(f)$, can be removed to some extent. Since we assume that the wireless channel only introduces attenuation, it only causes a scaling of the magnitude of the transfer function. Therefore, by using calibration values and scaling, an estimate of the tag transfer function, $\tilde{H}_k(f)$, can be obtained from the received signal $Y(f)$. Next, we will focus on applying the signal space representation to refine the detection of the information from this estimated tag transfer function $\tilde{H}_k(f)$.

2.3.1 Signal Space Representation of Frequency Signatures

In order to represent the RFID frequency signatures in a signal space, first, we need to find an orthonormal set of basis functions $\phi_i(f)$, $i = 1,...,N$ where any frequency signature, $H_k(f)$, can be written as a unique linear combination of $\phi_i(f)$s as in Equation 2.2. Here, as opposed to analyzing in time domain as in the case of the conventional communication systems, which was presented earlier, the analysis is done in the frequency domain. This is because the retransmission-based chipless RFID tag considered in this analysis represents information bits as abrupt changes in the frequency spectrum.

The first step in the derivation of the set of orthonormal basis functions is forming a matrix $\mathbf{H}$ that contains all the frequency signatures used for the RFID system. Let each tag frequency signature $H_k(f)$ be sampled in the frequency domain with m frequency samples, which results in an m-dimensional vector $\mathbf{h}_k$. Since $\mathbf{h}_k$ has both amplitude and phase information, it is generally a complex vector, that is, $\mathbf{h}_k \in \mathbb{C}^m$, where $\mathbb{C}^m$ is the set of m-dimensional complex vectors. The matrix $\mathbf{H}$ is formed by arranging the $n = 2^b$ distinct frequency signatures, $\mathbf{h}_k$s, as column vectors as follows:

$$\mathbf{H} = \begin{bmatrix} \mathbf{h}_1 & \mathbf{h}_2 & \cdots & \mathbf{h}_k & \cdots & \mathbf{h}_n \end{bmatrix} \tag{2.9}$$

Using singular value decomposition [12], the matrix $\mathbf{H}$ can be decomposed as follows:

$$\mathbf{H} = \Phi \Sigma \Theta^H \tag{2.10}$$

where Φ and Θ are unitary matrices composed of orthonormal column vectors $\boldsymbol{\varphi}_i$ and $\boldsymbol{\theta}_i$, respectively. The operator $(.)^H$ denotes the Hermitian transpose. The matrix Σ is a diagonal and positive definite matrix, which is called the singular value matrix containing singular values σ_i. The matrix is a full rank matrix since it has n mutually independent column vectors $\mathbf{h}_k$, $k = 1,\ldots,n$. Therefore, we can rewrite Equation 2.10 as

$$\underset{m \times n}{\mathbf{H} = \begin{bmatrix} \mathbf{h}_1 & \mathbf{h}_2 & \cdots & \mathbf{h}_k & \cdots & \mathbf{h}_n \end{bmatrix}} = \underset{m \times n}{\begin{bmatrix} \boldsymbol{\varphi}_1 & \boldsymbol{\varphi}_2 & \cdots & \cdots & \boldsymbol{\varphi}_n \end{bmatrix}} \underset{n \times n}{\begin{bmatrix} \sigma_1 & & & 0 \\ & \sigma_2 & & \\ & & \ddots & \\ 0 & & & \sigma_n \end{bmatrix}} \underset{n \times n}{\begin{bmatrix} \boldsymbol{\theta}_1^H \\ \boldsymbol{\theta}_2^H \\ \cdot \\ \cdot \\ \cdot \\ \boldsymbol{\theta}_n^H \end{bmatrix}}$$

$$\tag{2.11}$$

Since $\boldsymbol{\varphi}_i, i = 1,\ldots,n$ is a set of orthogonal column vectors, they serve as a good choice for a set of basis functions to represent the tag signatures $\mathbf{h}_k$. As discussed in the previous section, the amount of a particular $\boldsymbol{\varphi}_i$ contained in a vector $\mathbf{h}_k$ can be found by taking the inner product between them, that is, $\langle \mathbf{h}_k, \boldsymbol{\varphi}_i \rangle = \mathbf{h}_k^H \boldsymbol{\varphi}_i$. Further simplification of Equation 2.11 gives

$$
\begin{bmatrix} \mathbf{h}_1^H \\ \mathbf{h}_1^H \\ \vdots \\ \vdots \\ \mathbf{h}_n^H \end{bmatrix} \quad \varphi_i = \sigma_i \quad \theta_i = \sigma_i \begin{bmatrix} \theta_1^i \\ \theta_2^i \\ \theta_3^i \\ \vdots \\ \theta_n^i \end{bmatrix} \tag{2.12}
$$

$$
n \times m \quad m \times 1 \quad n \times 1 \quad n \times 1
$$

where θ_j^i is the jth element in the column vector $\boldsymbol{\theta}_j$. Therefore, from Equation 2.12, we see that the inner product between vectors $\mathbf{h}_k$ and $\boldsymbol{\varphi}_i$ can be expressed as

$$
\langle \mathbf{h}_k, \boldsymbol{\varphi}_i \rangle = \mathbf{h}_k^H \boldsymbol{\varphi}_i = \sigma_i \theta_k^i \tag{2.13}
$$

Hence, we can write

$$
\mathbf{h}_k = \sum_{i=1}^n \langle \mathbf{h}_k, \boldsymbol{\varphi}_i \rangle \boldsymbol{\varphi}_i = \sum_{i=1}^n \left(\sigma_i \theta_k^i \right) \boldsymbol{\varphi}_i \tag{2.14}
$$

Since only a small "L" number of σ_i are large and the rest are negligible, each $\mathbf{h}_k$ can be approximated using a few $\boldsymbol{\varphi}_i$ as expressed in Equation 2.15. This approach is commonly used in many applications involving model simplification [14]:

$$
\mathbf{h}_k \approx \sum_{i=1}^L \langle \mathbf{h}_k, \boldsymbol{\varphi}_i \rangle \boldsymbol{\varphi}_i = \sum_{i=1}^L \left(\sigma_i v_k^i \right) \boldsymbol{\varphi}_i \tag{2.15}
$$

Therefore, $\boldsymbol{\varphi}_i$, $i = 1,\ldots,L$ serve as an orthonormal basis for the tag frequency signatures $\mathbf{h}_k$. This will enable the formation of an L-dimensional signal space in which the 2^b tag signatures $\mathbf{h}_k$ are represented as 2^b signal points $\mathbf{s}_k \in \mathbb{C}^L$ (note that $\mathbf{s}_k \neq \mathbf{h}_k$ because $\mathbf{h}_k$s are vectors in an m-dimensional space, whereas $\mathbf{s}_k$s are vectors in an L-dimensional space). Each $\mathbf{s}_k$ corresponding to a tag signature $\mathbf{h}_k$ will have the following coordinates in the signal space:

$$
\mathbf{s}_k \equiv \left[\langle \mathbf{h}_k, \boldsymbol{\varphi}_1 \rangle, \langle \mathbf{h}_k, \boldsymbol{\varphi}_2 \rangle, \ldots, \langle \mathbf{h}_k, \boldsymbol{\varphi}_L \rangle \right] \tag{2.16}
$$

The signal points $\mathbf{s}_k, k = 1,\ldots,n$ serve as a signal constellation "C," which is a subset of the signal space. These $n = 2^b$ constellation points are used in the detection of information bits from the estimated tag signature $\tilde{H}_k(f)$.

Let $\tilde{\mathbf{h}}_k$ be the sampled estimate of the actual tag signature $H_k(f)$ that the reader estimates using the total received signal $Y(f)$. The structure of the detection algorithm is similar to that shown in Figure 2.4. The detection process involves in first calculating the inner products between $\tilde{\mathbf{h}}_k$ and each of the basis functions φ_i, $i = 1,\ldots,L$. The inner product coefficients obtained give an idea of how much of each basis function φ_i is contained in $\tilde{\mathbf{h}}_k$. These inner product coefficients, $r_i = \langle \tilde{\mathbf{h}}_k, \varphi_i \rangle$, form the noisy received signal point $\mathbf{r} = [\mathbf{r}_1,\ldots,\mathbf{r}_L]$ in $\mathbb{C}^L$. For the purpose of estimating the constellation point corresponding to $\tilde{\mathbf{h}}_k$, we use the maximum likelihood decoding concept [11]. Here, given the received vector and the noise statistics, the reader has to determine the most likely tag signal point $\hat{\mathbf{s}}$ that was responsible for causing the received signal point $\mathbf{r}$. For simplicity, we assume that the noise component has a Gaussian probability distribution where the maximum likelihood decoding process simplifies to minimum distance decoding [11]. Therefore, next, the Euclidian distance from each of the constellation points, $\mathbf{s}_k$, to the noisy received signal point, $\mathbf{r}$, is calculated. The constellation point that is closest to $\mathbf{r}$ is selected as the constellation point $\hat{\mathbf{s}}$ corresponding to the actual tag signature $H_k(f)$, which gives the information stored in the tag. The earlier explained procedure can be summarized by the following equation:

$$\hat{\mathbf{s}} = \arg\left(\min_{\mathbf{s}_i \in C}\left\{\|\mathbf{r} - \mathbf{s}_i\|^2\right\}\right) \qquad (2.17)$$

2.3.2 Application of New Model

We now discuss the application of the new mathematical model on simulation and measurement results obtained from a prototype chipless RFID system [9, 10]. For this purpose, a chipless RFID system with tags capable of carrying three bits of data was designed. A low number of bits were considered in the design for the purpose of conducting a comprehensive and thorough analysis. The tag essentially consists of two monopole antennas and a spiral resonator-based passive microwave filter. All components of the tag are based on coplanar waveguide (CPW) theory, which only demands a single surface of conductive material, due to the convenience of fabrication.

As explained in the previous sections, the data is represented in features present in the transfer function $H_k(f)$ of the tag. This $H_k(f)$ is essentially based on passive microwave filter design. By the use of different resonating structures, abrupt and sharp resonances or

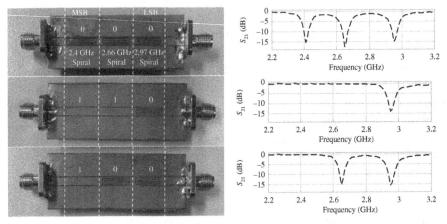

Figure 2.7 Spiral resonator-based filters carrying data "000," "110," and "100" and their respective transfer function $H_k(f)$. Each filter has the dimensions 2 cm × 5 cm and is fabricated on substrate Taconic TLX-0.

attenuations can be caused in $H_k(f)$. These resonances are used to represent logic "0" or "1" of the data bits carried by the chipless RFID tag. For the design of the 3-bit chipless RFID tag, a spiral resonator-based microwave filter was chosen. This filter consists of spiral resonators with each spiral resonating at a distinct frequency operating as a notch filter at the resonance. Eight different filters were designed using binary combinations of three different spiral resonators to produce the eight distinct $H_k(f)$s required to represent each of the 3-bit binary data combinations. Figure 2.7 shows three of the filters that are used in tags carrying data "000," "100," and "110."

Clearly, we can observe that by controlling the absence and presence of the spiral resonators, the frequency characteristics of the $H_k(f)$ can be directly changed, where a resonance is used to represent a data bit "0" and its absence represents a data bit "1." The filters were designed and simulated using the full-wave electromagnetic simulation software "Computer Simulation Technology (CST) Microwave Studio." The simulation results showed that the resonances occur at the frequencies 2.42, 2.66, and 2.97 GHz. The simulated and experimentally measured tag transfer functions $H_k(f)$ are shown in Figure 2.8. The measurements were done using a vector network analyzer. The forward transmission scatter parameter S_{21}, which is essentially equivalent to the filter transfer function $H_k(f)$, was measured using the instrument. The lowest and the highest resonance frequencies correspond to the most significant bit (MSB) and the least significant bit (LSB), respectively. From the CST

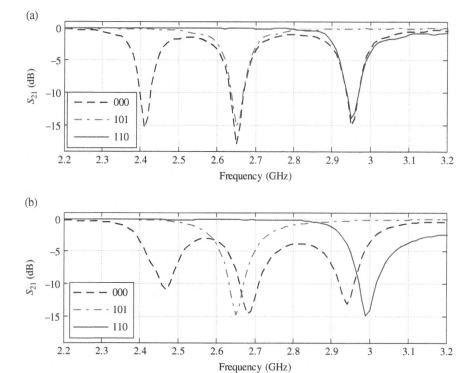

Figure 2.8 Tag transfer functions or frequency signatures, $H_k(f)$. (a) Simulated. (b) Measured.

simulation results shown in Figure 2.8a, it is clear that the resonances corresponding to the different data-carrying $H_k(f)$s accurately coincide to three distinct resonance frequencies, namely, 2.42, 2.66, and 2.96 GHz. However, the same cannot be said for the experimentally measured transfer functions of the different filters. The resonances vary from one filter to another with a tolerance of around ±50 MHz. These discrepancies are due to the minute fabrication errors that have caused the physical dimensions of the resonating spiral structure and the depth of the substrate material to vary from one filter to another. These deviations are ultimately seen as a form of noise in the detection process, and hence, it will result in an increase in detection error probability. From the simulation and experimental measurements, we can observe that the 3 dB bandwidth of the resonances is around 150–200 MHz.

Since the simulation results are close to the ideal theoretically expected results for $H_k(f)$, they were used as the reference transfer functions in calculating the set of basis functions discussed previously.

Table 2.1a shows the singular values resulting from performing singular value decomposition on the frequency signatures as expressed in Equation 2.10. Clearly, only the first four singular values are significant. Hence, all the $H_k(f)$s can be adequately described using only $L = 4$ basis functions φ_i, $i = 1, \ldots, L = 4$. Table 2.1b lists the inner product coefficients, $\langle \mathbf{h}_k, \varphi_i \rangle$, between different tag signatures and basis functions φ_i, $i = 1, \ldots, L = 4$. The inner product $\langle \mathbf{h}_k, \varphi_i \rangle$ is a measure of the amount of basis function φ_i in the signature $\mathbf{h}_k$. According to the table, it is clear that φ_1 is the largest component in all the $\mathbf{h}_k$s. However, the relative variation of $\langle \mathbf{h}_k, \varphi_i \rangle$ from one signature to another is small. Therefore, in identifying the individual $\mathbf{h}_k$s, φ_1 provides little information. As opposed to φ_1, the other basis functions show larger relative variations among different tag signatures. Therefore, the basis functions φ_2, φ_3, and φ_4 are sufficient to uniquely identify each signature. Hence, using these basis functions, a three-dimensional signal space was constructed where all the eight frequency signatures corresponding to data "000," "001," "010," …, "111" were plotted as signal points (constellations points) in it as shown in Figure 2.9. These eight signal points can be considered a constellation against which measured unknown tag signatures will be matched in order to extract their data.

Figures 2.8 and 2.9 show the clear contrast between the two representations (frequency domain and signal space) of $H_k(f)$. In the former, the tag transfer functions are represented with respect to frequency where

Table 2.1   (a) Singular Values. (b) Inner Product Coefficients [9]

(a)							
σ_1	σ_2	σ_3	σ_4	σ_5	σ_6	σ_7	σ_8
31.54	2.75	2.58	2.29	0.23	0.18	0.12	0.06

(b)			
	$\langle \mathbf{h}_k, \varphi_1 \rangle$	$\langle \mathbf{h}_k, \varphi_2 \rangle$	$\langle \mathbf{h}_k, \varphi_3 \rangle$	$\langle \mathbf{h}_k, \varphi_4 \rangle$
000 $\mathbf{h}_1$	10.29	0.43	−0.59	1.33
001 $\mathbf{h}_2$	10.94	1.69	0.09	0.19
010 $\mathbf{h}_3$	10.76	−0.75	0.91	0.98
011 $\mathbf{h}_4$	11.37	0.34	1.51	−0.02
100 $\mathbf{h}_5$	10.83	−0.54	−1.52	0.17
101 $\mathbf{h}_6$	11.52	0.86	−0.78	−0.95
110 $\mathbf{h}_7$	11.34	−1.63	−0.19	−0.19
111 $\mathbf{h}_8$	12.05	−0.36	0.49	−1.23

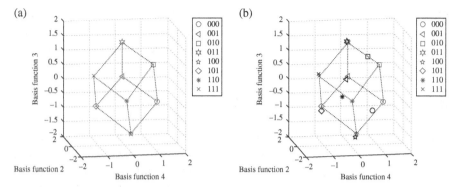

Figure 2.9 Signal space representation of the information-carrying transfer functions $H_k(f)$ [9]. (a) Simulated. (b) Measured.

the resonances are clearly visible at certain frequencies where as in the latter, the tag transfer functions are represented as points in a three-dimensional space where the Euclidian distance separating them defines the uniqueness of each point or tag transfer function. In a way, the representation of transfer functions in frequency domain can also be thought of as a signal space representation where there exist an infinite number of basis functions in which the basis functions are the individual complex exponential frequencies. In such a representation, each frequency in the frequency axis can be thought of as a basis function, which is also a distinct dimension of the corresponding infinite-dimensional signal space. Through the singular value decomposition-based approach, an alternative basis for representing these transfer functions is realized, and of those, only the most essential basis functions are chosen to represent the transfer functions. This gives rise to the proposed method where the $H_k(f)$s are represented using a smaller finite-dimensional signal space. In Figure 2.9, the experimentally measured tag transfer functions and the simultated tag transfer functions are shown . The discrepancies observed in the resonance frequencies of the measured and simultated $H_k(f)$ seen in Figure 2.9 are reflected as displacements from the expected locations (locations of the corresponding constellation points) in the signal space. As the error in the frequency-domain representation gets larger, the displacements observed in the signal space becomes greater. Therefore, we can see that minute fabrication errors can be interpreted as a noise source that causes the signal points to be displaced from their intended locations. Despite this noise, the signal points corresponding to the measured tag transfer functions shown in Figure 2.9b can be detected accurately. This is because the noise introduced by the fabrication error is not large enough for them to be displaced significantly and be mistaken for a wrong constellation point, that is, the minimum distance defined by Equation 2.17 is still achieved with the correct constellation point.

2.4   PERFORMANCE ANALYSIS

This section presents a discussion and an analysis on how the detection performs under additive noise. Noise affecting the system is generated from two main noise sources: the additive thermal noise introduced at the receiver and the noise introduced due to small fabrication errors due to the finite precision in manufacturing a tag. For the ease of analysis, we consider the combined effect of the total noise affecting the system to be additive white and Gaussian. AWGN causes the received signal point $\mathbf{r}$ to be displaced from the location of the correct constellation point $\mathbf{s}_k$ and causes detection errors. The amount of AWGN affecting the data-carrying signal is usually defined using a metric known as signal-to-noise ratio (SNR) in conventional communication systems. However, in the context of chipless RFID tag signatures, the definition of SNR cannot be directly used since the data is not contained in the received signal but in the transfer function of the tag filter. Therefore, in order to quantify the amount of noise affecting the tag signature, the following quality factor γ is defined [9]:

$$\gamma = \frac{d_0^2}{4\sigma^2} \tag{2.18}$$

where d_0 is the average length between adjacent constellation points (average length of the edges of the constellation) and σ^2 is the noise power spectral density. From Equation 2.18, we can see that both the distance between adjacent constellation points and the noise affecting them are important in defining the overall performance. Figure 2.10 shows the effect of AWGN on the received signal point $\mathbf{r}$. The figure shows the results of a simulation on detecting the "110" data-carrying tag in 100 trials. We can clearly see that a cloud is formed by the 100 noisy received signal points centering around the "110" constellation point. When the noise affecting the system increases, this cloud gets larger and spreads closer to the other constellation points, which causes an increase in the probability of detection error [11].

Figure 2.11 shows the simulated performance of the detection using minimum distance through the signal space representation.

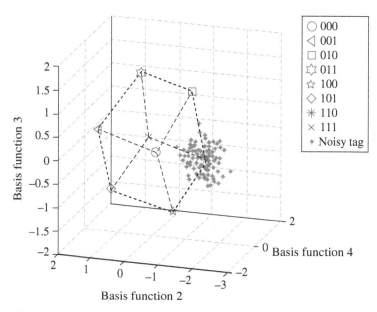

Figure 2.10   The noise cloud formed by 100 received signal points for $\gamma = 10$ dB, [9].

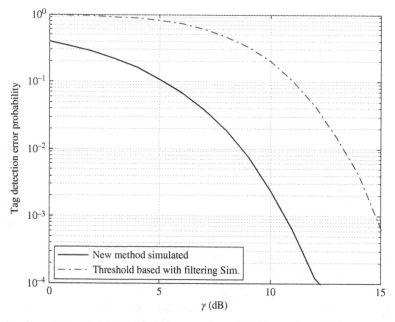

Figure 2.11   Tag detection error probability against γ, [9].

The probability of detection error is plotted against different noise levels that are defined using the metric γ. These results were obtained using Monte Carlo simulations for 100,000 trials. The experiment involves the detection of a randomly chosen tag (out of the eight possible tags) in AWGN. Figure 2.11 also shows the detection performance when a threshold-level-based detection scheme is used. In this detection method, the signal is first filtered to reduce the effects of noise on the signal. Afterwards, the filtered signal level near the resonance frequencies ($\pm100\,$MHz) was compared against a predefined fixed threshold level in order to determine whether a "0" or "1" bit is present. If signal level is below, the threshold "0" is detected; if not, "1" is detected. From the figure, it is clear that the minimum distance-based detection has better performance over frequency-domain fixed threshold-based detection in AWGN.

2.5   EXPERIMENTAL RESULTS USING THE COMPLETE TAG

The performance analysis simulations and discussions presented so far dealt only with the tag transfer function, $H_k(f)$, where it was assumed that it can be estimated with reasonable accuracy. These analyses did not include the effects of signal transmission through the wireless channel. As expressed in Equation 2.8, the accurate estimation of $H_k(f)$ relies on successful removal and mitigation of the undesirable effects caused by the wireless channel and the other interferences. In order to test the new detection under all these effects, a complete chipless tag was constructed, and experiments were performed.

The complete retransmission-based chipless RFID tag used for the experiment is shown in Figure 2.12a. The tag consists of two monopole antennas, one for receiving and the other for retransmitting, and a spiral resonator filter. The two monopole antennas are oriented perpendicular to each in order to achieve orthogonal polarities in the received and transmitted signals. This would help to minimize the interference between the two antennas of the tag. Figure 2.12b shows the experimental setup used for reading the chipless tags. A two-port vector network analyzer (VNA) was used for reading the tag where two high gain log periodic dipole antennas were attached to the two ports of the VNA. Here, port 1 of the VNA serves as the transmitting port, and port 2 of VNA serves as the receiving port. The interrogation signal transmitted by the VNA is a chirp signal with a start frequency of 2.2 GHz and stop frequency of 3.2 GHz. The forward transmission

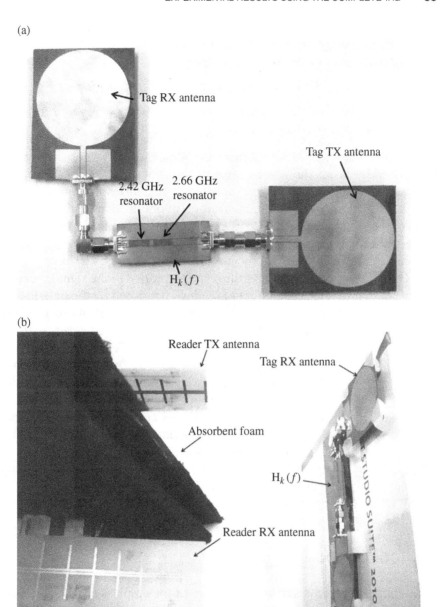

Figure 2.12 (a) Complete retransmission-based chipless RFID tag. (b) Experimental setup for taking tag measurements [9].

coefficient, S_{21}, serves as the raw tag measurement. From the definition of S_{21}, we can write

$$S_{21} = \frac{Y(f)}{X(f)} = W_F(f)H_k(f)W_R(f) + I(f) + N(f) \qquad (2.19)$$

The measured S_{21} parameters of a tag, which carries data "100," for different distances 1, 5, and 10 cm away from the reader, are shown in Figure 2.13a. From the results, we can see that only the measurement done at 1 cm shows a clear resemblance to the corresponding tag filter transfer function $H_5(f)$. In the measurement obtained at 5 cm, the sharpness of the resonance occurring near 2.96 GHz has reduced. At 10 cm, the observed tag frequency signature has become more distorted, that is, the resonance that is supposed to occur at 2.96 GHz appears to have shifted to 3 GHz. These frequency signatures can be plotted in the signal space as shown in Figure 2.13b. The effect of scaling introduced by the path loss of the wireless channel for different distances can be neutralized to some extent by normalizing the measured S_{21} parameters. However, without using a reference or calibration measurement, the effect of the unwanted interference is difficult to remove. We hypothesize that this interference is one of the reasons for the distortion that appears in the frequency signature at distance beyond 5 cm. From the signal space shown in Figure 2.13b, we can see that the estimates of $H_5(f)$ plotted in the signal space can be correctly detected as carrying data "100" for distances 1 and 5 cm. However, due to the presence of greater distortion, the estimate of $H_5(f)$ obtained from the measurement done at 10 cm is not correctly detected to be carrying data "100." It is erroneously detected to be the "000" data-carrying constellation point hence causing a detection error.

2.6   CONCLUSION

This chapter presents a new perspective of the chipless RFID detection problem where the chipless RFID tag signatures are geometrically represented in a signal space. The operation of the chipless RFID system is compared and contrasted with the operation of a conventional digital communication system. A brief review of signal space representation in the context of a digital communication system is provided. A novel method for detecting data in the frequency signatures of chipless RFID tags is formulated based on the representation of tag frequency signatures as signal points in a signal space. Spiral resonator-based

(a)

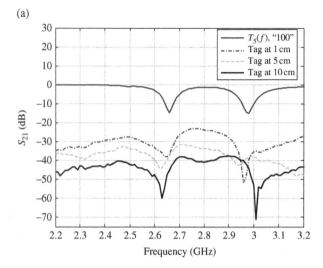

(b)

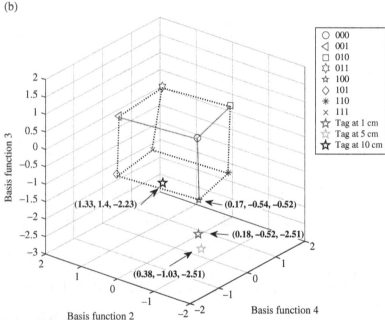

Figure 2.13 (a) The reference tag transfer function for data "100" and the frequency signatures of "100" tag measured at different distances. (b) Signal space representation of estimated tag frequency signature [9].

chipless RFID tags are used in order to validate the new approach. Monte Carlo simulations show that the proposed detection method has better detection performance compared to threshold-based detection of bits in frequency domain.

REFERENCES

1. A. Goldsmith, *Wireless Communications* New York: Cambridge University Press, 2005.
2. S. Sesia, et al., *LTE, The UMTS Long Term Evolution: From Theory to Practice*, Chichester: John Wiley & Sons, Ltd, 2009.
3. S. Preradovic and N. C. Karmakar, "Chipless RFID: bar code of the future," *IEEE Microwave Magazine*, vol. 11, pp. 87–97, 2010.
4. S. Preradovic, I. Balbin, N. C. Karmakar, and G. Swiegers, "*A novel chipless RFID system based on planar multiresonators for barcode replacement*," presented at the 2000 IEEE International Conference on RFID, Las Vegas, Nevada, April 16–17, 2008.
5. S. Preradovic and N. C. Karmakar, "*Design of fully printable planar chipless RFID transponder with 35-bit data capacity*," presented at the European Microwave Conference, 2009. EuMC 2009, Rome, September 29–October 1, 2009.
6. A. Chamarti and K. Varahramyan, "Transmission delay line based ID generation circuit for RFID applications," *IEEE Microwave and Wireless Component Letters*, vol. 16, pp. 588–590, 2006.
7. S. Hu, C. L. Law, and W. Dou, "A balloon-shaped monopole antenna for passive UWB-RFID tag applications," *IEEE Antennas and Wireless Propagation Letters*, vol. 7, pp. 366–368, 2008.
8. B. Shao, Q. Chen, Y. Amin, S. M. David, R. Liu, and L. R. Zheng, "*An ultra-low-cost RFID tag with 1.67 Gbps data rate by ink-jet printing on paper substrate*," presented at the IEEE Asian Solid-State Circuits Conference (A-SSCC), Beijing, 2010.
9. P. Kalansuriya, N. C. Karmakar, and E. Viterbo, "A novel approach in the detection of chipless RFID," in *Chipless and Conventional Radio Frequency Identification: Systems for Ubiquitous Tagging*, N. Karmakar, Ed., Hershey, PA: IGI Global, 2012, pp. 218–233.
10. P. Kalansuriya, N. C. Karmakar, and E. Viterbo, "*Signal space representation of chipless RFID tag frequency signatures*," presented at the IEEE Global Communications Conference, GLOBECOM 2011, Houston, TX, December 5–9, 2011.
11. S. Haykin, *Communication Systems*, 5 ed., New York: John Wiley & Sons, Inc., 2009.
12. G. Strang, *Linear Algebra and Its Applications*, 4 ed., Wellesley: Wellesley-Cambridge Press, 2009.
13. S. Preradovic and N. C. Karmakar, "*Design of short range chipless RFID reader prototype*," presented at the 2009 5th International Conference on Intelligent Sensors, Sensor Networks and Information Processing (ISSNIP), Melbourne, Australia, December 7–10, 2009.
14. S. Boyd. (2007, December 1, 2010). *Lecture 16 SVD applications*. Available: http://see.stanford.edu/materials/lsoeldsee263/16-svd.pdf (accessed July 7, 2015).

CHAPTER 3

TIME-DOMAIN ANALYSIS OF FREQUENCY SIGNATURE-BASED CHIPLESS RFID

3.1 LIMITATIONS OF CURRENT CONTINUOUS-WAVE SWEPT FREQUENCY INTERROGATION AND READING METHODS FOR CHIPLESS RFID

This section describes the frequency swept continuous-wave approach used for reading chipless RFID tags [1]. Here, the approach of reading tags is summarized (refer Chapter 2 for more details) and the limitations and challenges faced by this approach of reading chipless RFID tags are discussed.

In a typical frequency-swept reader, a voltage-controlled oscillator (VCO) is used to generate continuous radio frequency waves to inter-rogate the chipless RFID at specific frequencies, which are then transmitted by an antenna. By sweeping the voltage supplied to the VCO, the frequency can be swept from a specific start frequency to a specific stop frequency. A receiving antenna picks up the signals from the tag where the amplitude and the phase are inspected to deter-mine the data encoded in the tag. Here, the antennas are designed so that there is a good isolation between them to minimize the coupling

Chipless Radio Frequency Identification Reader Signal Processing, First Edition.
Nemai Chandra Karmakar, Prasanna Kalansuriya, Rubayet E. Azim and Randika Koswatta.
© 2016 John Wiley & Sons, Inc. Published 2016 by John Wiley & Sons, Inc.

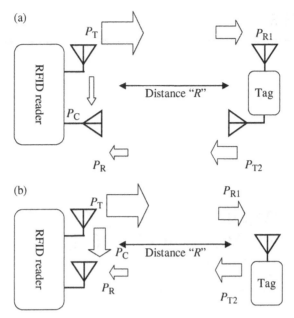

Figure 3.1 (a) A retransmission-based chipless RFID system. (b) A backscatter-based chipless RFID system.

between transmit and receive antennas. However, even with a very good level of isolation, the tag will only perform well up to a short range beyond which accurate calibration, using reference measurements, is required in order to estimate the data encoded in the tag.

Figure 3.1 shows two types of frequency signature-based chipless RFID systems. First, let's consider the retransmission-based chipless RFID system [2] shown in Figure 3.1a. Let P_T be the power being transmitted by the RFID reader and P_{R1} be the power received at the RFID tag antenna, where the tag is at a distance "R" from the reader. Assuming that the path through the tag is lossless, the retransmitted power P_{T2} by the tag will be equal to the power received, P_{R1}. Using Friis transmission equation [3], we can write an expression for the received power at the tag as

$$P_{R1} = P_T \left(\frac{\lambda}{4\pi R} \right)^2 G_1 G_2 \tag{3.1}$$

where G_1 is the gain of the antennas at the reader and G_2 is the gain of the antennas at the tag. Since $P_{T2} = P_{R1}$, we can express the power received at the reader P_R to be

Table 3.1 Power Received at the Reader Due to the Signals Retransmitted by the Chipless RFID Tag

Distance R (cm)	Power received P_R	Ratio of Tag Power to Coupled Power for Different Levels of Isolation		
		40 dB	50 dB	60 dB
5	$1.8 \times 10^{-2} P_T$	180	1,800	18,000
10	$1.1 \times 10^{-3} P_T$	11	110	1,100
15	$2.2 \times 10^{-4} P_T$	2.2	22	220
20	$7.1 \times 10^{-5} P_T$	0.71	7.1	71
25	$2.9 \times 10^{-5} P_T$	0.29	2.9	29
30	$1.4 \times 10^{-5} P_T$	0.14	1.4	14
35	$7.6 \times 10^{-6} P_T$	0.076	0.76	7.6
40	$4.4 \times 10^{-6} P_T$	0.044	0.44	4.4

Shaded area of the table shows the situations where the data in the tag cannot be accurately detected.

$$P_R = P_{T2} \left(\frac{\lambda}{4\pi R} \right)^2 G_1 G_2 = P_T \left(\frac{\lambda}{4\pi R} \right)^4 (G_1 G_2)^2 \qquad (3.2)$$

For a retransmission-based system, usually the transmit and receiving antennas are cross polarized in order to minimize the coupling between them. Let's assume that the isolation between the two antennas is around 40 dB, which is a typically good level of isolation. Therefore, the power coupled to the receiving antenna is $P_c = P_T \times 10^{-4}$. Consider the transmission of a continuous wave having a frequency of 5.8 GHz. Let's assume that the antennas at the reader have a gain of 10 dB and the tag antennas have a gain of 3 dB. Table 3.1 shows the calculated received power at the reader, P_R, as a fraction of the transmitted power, P_T, when the tag is placed at different distances from the reader. It is clear from these values that only a very small amount of power arrives back at the reader even when we assume ideal conditions where the path through the tag is lossless. Also, it is observed that when an isolation of 40 dB is achieved at the reader antennas with the tag at 15 cm from the reader, the coupled power, P_c, is comparable to power received from the tag P_R. When these two signals add constructively or destructively depending on their phase, which manifests as a spectral peak or null that can be falsely perceived as a resonance of the tag, and consequently, detection errors can occur. When the coupled power level is larger than the received power from the tag, it is not possible to detect the data contained in the signal because the coupled signal simply overwhelms and drowns the signal received from the tag. The range can be improved a little by improving the isolation between the two antennas, where an isolation of 50 dB allows the tag to be read accurately up to 15 cm. The read range cannot be improved by simply increasing transmit power, P_T, because the coupled power also proportionally increases and the ratio of received tag power to couple power remains the same.

In order to read the tag beyond this limiting distance, some form of calibration is required where the effect of the coupling is removed. In the work of Ref. [1], the authors use a calibration tag that helps to identify the changes of the amplitude or phase due to the effect of the chipless tag only. However, the use of a calibration tag limits the use of the system in applications that does not maintain a fixed distance to the tag and in situations where the environment is highly dynamic. Another method to remove the effect of the coupling signal is to perform a tag-free measurement and obtain the amplitude and phase of the coupled signal alone and perform a vector subtraction for all the subsequent measurement with the tag. However, due to the presence of the tag in the near field environment, the coupling will be slightly different than in a completely tag-free situation, which would again affect the accuracy.

Apart from removing the coupled power component, the accuracy of the detection can be improved by enhancing the signal coming back from the tag. For this, the gains of the transmit and receive antennas at the reader and the tag can be improved using better antenna design techniques. By placing absorber materials and by increasing the distance between the receive and transmit antennas, the isolation between them can be further improved to enhance the range. However, it is clear from the values in Table 3.1 that even with an isolation of 60 dB between the antennas, the maximum theoretical range that can be achieved without-calibration is around 30–35 cm. Also, the process of improving isolation would render the RFID reader unit to be bulky and cumbersome to handle.

In the case of backscatter-based chipless RFID systems [4] shown in Figure 3.1b, where the antennas at the reader are copolar, the effect of the coupled signal is more severe. This is because the isolation between the antennas is poorer in the copolar orientation. Apart from the coupling, another factor needs to be addressed in backscatter-based chipless RFID systems. That is, when the total backscattered signal from the tag is considered, only a small portion of it contributes to conveying the information stored in the chipless RFID tag where majority of the backscattered energy does not contain any encoded information in the tag. It is shown in Refs. [5, 6] that the information is contained in the *antenna mode* backscatter or the backscatter that results due to the resonant properties of the tag. When the signal is viewed only in the frequency domain, it is difficult to differentiate between the information-carrying *antenna mode* backscatter and the other backscattered signals. Through tag design, it is possible to maximize the effect of the *antenna mode* backscatter to make it more prominent amidst the other signals as in Ref. [7]. However, it is not possible to completely separate only the information-carrying component of the backscatter and view its spectral content, when viewed in the frequency domain.

3.2 UWB-IR INTERROGATION OF TIME-DOMAIN REFLECTOMETRY-BASED CHIPLESS RFID

In this section, we explain the use of ultra-wideband (UWB) impulses in reading time-domain reflectometry (TDR)-based chipless tags. This will provide a clear contrast between the clarity and the detail to which the received backscatter can be perceived and analyzed in the time domain as opposed to the frequency-domain perception, which was described previously. Here, it is clearly possible to distinctly observe and separate out the different components making up the total signal arriving at the reader without the need of calibration tags. A brief summary of the operation will be presented focusing more on the key components that make up the back scatter than the operation of the tag itself.

Figure 3.2 shows a typical TDR-based chipless RFID system [8, 9]. The RFID reader is essentially an ultra-wide bandwidth impulse radar (UWB-IR) system. It transmits a UWB pulse and listens to the reflections coming back from the tag. The typical TDR chipless RFID tag [9] consists of a broadband antenna and a transmission line segment, which is terminated with a load. The length of this transmission line segment is used to encode data in TDR-based chipless RFID tags. Before presenting a detailed explanation about the electromagnetic interaction that occurs at the tag, it is convenient to clearly define two types of backscatter. When electromagnetic radiation is incident on an antenna, the surface currents induced on the surface of the antenna give rise to an initial backscatter. This initial backscatter depends only on the shape, size, and material properties of the antenna and is called the *structural mode* backscatter [3, 8, 10]. Another portion of the incident

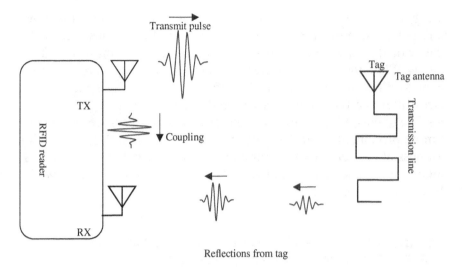

Figure 3.2 Time-domain reflectometry-based chipless RFID system.

electromagnetic energy is harnessed and received by the antenna. Depending on the matching condition with the load connected to the antenna at its termination, this received energy will be reflected and reradiated giving rise to a secondary backscatter called the *antenna mode* backscatter [3, 8, 10]. This secondary backscatter is dependent upon the radiation characteristics of the antenna and the reflection coefficient at its termination. The *antenna mode* backscatter will be ideally zero when the load connected to the antenna is well matched.

Figure 3.3 shows the total received signal at the RFID reader when the tag is located 45 cm away, which was obtained through full-wave simulation using CST Microwave Studio™. Here, the transmit antenna and receive antenna are of same polarity. The received signal consists of three parts. The first and largest component received is the coupling signal. Due to the proximity of the received antenna to the transmit antenna, a strong signal is coupled to the receiving antenna. When the transmit pulse reaches the chipless RFID tag, it interacts with the tag antenna and gives rise to a *structural mode* backscatter, which is the next largest signal received at the reader. Finally, the *antenna mode* backscatter, which comprises of the energy that was received by the tag antenna, is received at the reader. The *antenna mode* backscatter is the weakest among these three components making up the total received signal. There is a clear separation between the *structural mode* backscatter and the *antenna mode* backscatter due to the use of transmission line segment. TDR-based chipless RFID tags makes use of this delay

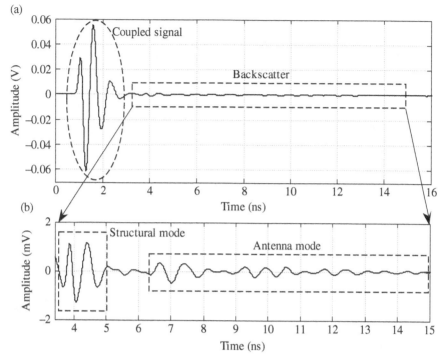

Figure 3.3 Total received signal at the RFID reader. (a) Total signal. (b) Enlarged portion of the signal that contains the *structural mode* and *antenna mode* backscatter.

between these two modes to encode information in the tag. In Figure 3.3a and b, all the components making up the received signal can be clearly and distinctly observed. The frequency-domain representation of the exact same signal, obtained through the discrete Fourier transform (DFT), is shown in Figure 3.4 where none of these key components making up the total received signal can be clearly distinguished. Therefore, it is evident that by perceiving the signal in the time domain, it is possible to separate the key components of the received signal, which gives the ability to extract the information-carrying component of the received signal without using calibration tag measurements.

From the time-domain signals provided in Figure 3.3a and b, it is possible to compute the amount of energy contained in each of the components making up the total signal. The energy contained in the time-domain signal for a time duration lasting from $t = t_1$ to $t = t_2$ can be expressed as

$$E = \frac{1}{R}\int_{t_1}^{t_2}\{v(t)\}^2\,dt \qquad (3.3)$$

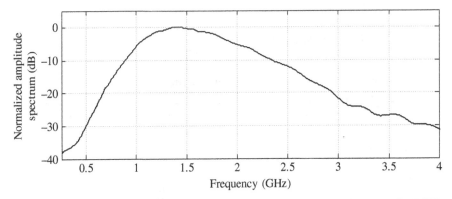

Figure 3.4   Normalized amplitude spectrum of the total received signal at the RFID reader.

Table 3.2   The Amount of Energy Contained in Each of the Major Components Making up the Total Backscattered Signal

Component	Energy Contained (Assuming a of Load $R = 1\,\Omega$), pJ
UWB pulse injected to antenna (not shown)	151.34
Coupled signal (Fig. 3.3a)	1.323
Structural mode of backscatter (Fig. 3.3b)	7.1×10^{-4}
Antenna mode of backscatter (Fig. 3.3b)	1.6×10^{-4}

Table 3.2 shows the amounts of energy contained in each of the major components identified in the time-domain signal shown in Figure 3.3. The calculations performed using Equation 3.3 reveal that more than 99.99% of the energy in the total received signal is contained in the coupled signal. Also, when the energy of the backscatter is considered, more than 80% of the energy is contained in the *structural mode*. By careful design of the tag antenna, the *structural mode* backscatter can be minimized as in Ref. [8]. However, still, the *structural mode* will contribute to a major portion of the backscattered energy. In the frequency domain, the observed signal is a superposition of the frequency spectra of all these individual components where the effect of each individual component by itself is not visible. Hence, in order to improve the quality of the detection performed at the reader, the need for isolating each of these components in the time domain and analyzing them separately is apparent.

3.3 TIME-DOMAIN ANALYSIS OF FREQUENCY SIGNATURE-BASED CHIPLESS RFID

The details presented in the previous sections establish the need for analyzing the backscatter in the time domain. In this section, the procedure involved in time-domain analysis and signal processing is introduced. The benefits and advantages of time-domain analysis are also discussed.

Figure 3.5 illustrates the flow chart of the process involved in analysis of the backscatter. For the purpose of interrogating the chipless RFID tag, either a continuous wave frequency sweeping reader or UWB-IR can be used. When continuous wave interrogation is utilized, the time-domain signal is obtained by performing an inverse Fourier transform (FT) on the sampled frequency-domain amplitude and phase information. With a UWB-IR-based system, digital sampling will directly provide the required time-domain signal. Afterwards, the key signal components can be separated using a suitable windowing function. Also, denoising techniques such as wavelet-based denoising can be utilized to further enhance the signal-to-noise ratio (SNR) of the received backscatter. Once the information-carrying portion of the backscatter is windowed out, its spectral content can be analyzed using Fourier analysis, which will reveal the information that was encoded by the chipless tag in the frequency domain (frequency signature of the tag).

It is reported in the literature [9] that the typical TDR-based chipless RFID tag can be read from several meters away using UWB-IR techniques. Also, they require minimal calibration and do not require the use of calibration tags to remove unwanted interfering signals to enhance the backscatter component of the total received signal. By employing the same time-domain analysis and processing techniques on the backscatter from the frequency-domain chipless RFID tags, it will be possible to improve their range of detection without using calibration tags and reference measurements.

The next two sections will provide a detailed analysis of the backscatter using two specific types of chipless RFID tags as examples.

3.4 ANALYSIS OF BACKSCATTER FROM A MULTIRESONATOR LOADED CHIPLESS TAG

This section presents the analysis of backscatter from a multiresonator loaded chipless RFID tag [5]. First, the system model of the chipless RFID system is presented. A model for the chipless RFID tag is given

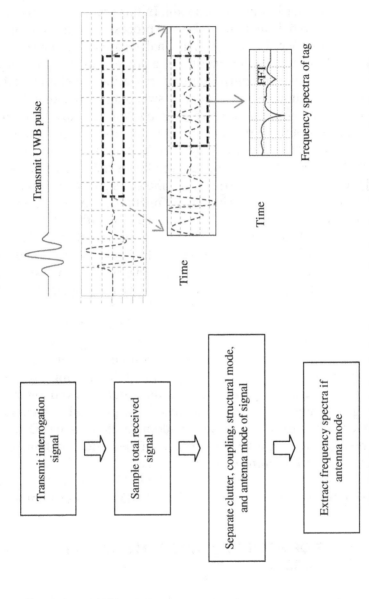

Figure 3.5 The flow of analyzing the backscatter and estimating the frequency signature of the tag.

and the resulting backscatter due to the reader interrogation signal is mathematically expressed. The constituents of the total received signal are clearly differentiated. The information-carrying component of the backscatter is identified, and an approach of estimating the tag frequency signature from the backscatter is discussed.

3.4.1   System Description and Mathematical Model for Backscatter Analysis

Figure 3.6 depicts the system model of the chipless RFID system considered in this analysis. The RFID reader consists of two antennas, one for transmitting the UWB interrogation pulse and the other for receiving the backscatter. The antennas are oriented in the copolar to each other since the tag backscatter is also expected to arrive with the same polarity. The chipless RFID tag consists of three parts: a broadband antenna for receiving the interrogation signal, a multiresonator planar filter that encodes information into the received signal, and a meandered transmission line that introduces a controlled delay in order to clearly differentiate the *antenna mode* and *structural mode* of the backscatter arriving at the reader.

Let $x(t)$ be the UWB pulse being transmitted by the RFID reader and let $y(t)$ be the total received signal at the reader. As explained in the previous sections, the first and the strongest component received is

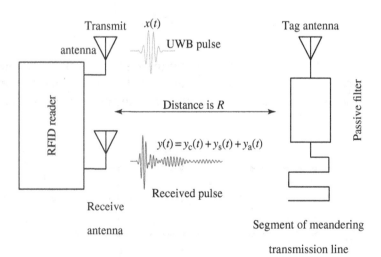

Figure 3.6   Chipless RFID system for reading multiresonator loaded chipless RFID tag [5].

the coupling signal $y_c(t)$. When $x(t)$ reaches the tag and interacts with the tag antenna and the metallic structure of the tag, it gives rise to a *structural mode* backscatter $y_s(t)$, which is followed by the *antenna mode* backscatter $y_a(t)$, after the delay introduced by the meandered transmission line. The *structural mode* results from surface reflection of the electromagnetic wave illuminating the chipless tag. On the other hand, the *antenna mode* results from the energy that is captured by the tag antenna. This energy passes through the multiresonator-based filter, which modifies and transforms the spectral content of this captured energy in order to encode the information contained in the chipless RFID tag. Afterwards, this transformed signal travels through the meandered transmission line and encounters an abrupt mismatch condition at the termination of the transmission line where it is reflected back toward the direction of the tag antenna and ultimately gets retransmitted. This retransmission forms the *antenna mode* backscatter. Here, the *antenna mode* backscatter lags the *structural mode* backscatter by the propagation delay introduced by the meandered transmission line.

The two-port network model of the chipless RFID tag is shown in Figure 3.7. The waves a and b represent the incident and reflected electromagnetic waves, respectively, where they are normalized to the free-space impedance [9]. The overall reflection coefficient presented by the tag to the wave incident on it can be expressed as [5]

$$\Gamma_{in} = S_{1,1}^a + \frac{S_{1,2}^a S_{2,1}^a \Gamma_f}{1 - S_{2,2}^a \Gamma_f}, \quad \Gamma_f = S_{1,1}^f + \frac{S_{1,2}^f S_{2,1}^f \Gamma_{out}}{1 - S_{2,2}^f \Gamma_{out}} \tag{3.4}$$

where $S_{i,j}^a$ and $S_{i,j}^f$ are the scattering parameters of the tag antenna and the multiresonator-based filter, respectively. Γ_f and Γ_{out} are the reflection coefficients seen at the input and the output port of the filter as

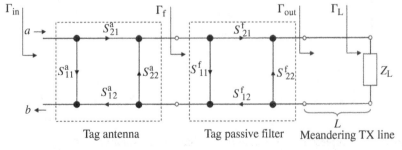

Figure 3.7   Two-port network model of the chipless RFID tag [5].

shown in Figure 3.7. When $S_{i,j}^{a}$ is considered, it should be noted that, here, $S_{1,1}^{a}$ does not correspond to the return loss profile of the tag antenna. It simply expresses the ratio of the reflected wave to the wave incident on the tag surface where the reflected portion gives rise to the *structural mode* of the backscatter. On the other hand, the scatter parameter $S_{2,2}^{a}$ corresponds to the return loss of the tag antenna, where it represents the ratio between the rejected and incident signals that arrive at the antenna terminals due to the reflection occurring at the mismatched load (refer Fig. 3.7). Further simplification of Equation 3.4 leads to the approximation

$$\Gamma_{in} \approx S_{1,1}^{a} + S_{2,1}^{a}S_{1,2}^{a}S_{1,1}^{f} + S_{1,2}^{a}S_{2,1}^{a}S_{2,1}^{f}S_{1,2}^{f}\Gamma_{out} \tag{3.5}$$

Let's assume that the meandered transmission line is a low-loss transmission line where the reflection coefficient, Γ_{out}, can be expressed as

$$\Gamma_{out} = \Gamma_{L}\exp(-j2\beta L) = \Gamma_{L}\exp(-j\omega\tau_{0}) \tag{3.6}$$

where L is the length of the transmission line segment, Γ_{L} is the reflection coefficient at the load, β is the phase constant, and $\tau_{0} = \frac{2L}{v}$ is the propagation delay due to the round-trip propagation of the wave along the meandering transmission line with v being the velocity of wave propagation. In this analysis, the transmission line is terminated using an open-circuit or short-circuit load, which makes the load reflection coefficient, Γ_{L}, to be either 1 or −1. By taking the inverse FT of Equation 3.5, the reflection coefficient can be expressed in the time domain as [5]

$$\Gamma_{in}(t) \approx S_{1,1}^{a}(t) + S_{2,1}^{a}(t)*S_{1,2}^{a}(t)*S_{1,1}^{f}(t)$$
$$+\Gamma_{L}*S_{2,1}^{a}(t)*S_{1,2}^{a}(t)*S_{2,1}^{f}(t)*S_{1,2}^{f}(t)*\delta(t-\tau_{0}) \tag{3.7}$$
$$\Gamma_{in}(t) \approx S_{1,1}^{a}(t) + p_{1}(t) + \Gamma_{L}p_{2}(t-\tau_{0})$$

where $p_{1}(t) = S_{2,1}^{a}(t)*S_{1,2}^{a}(t)*S_{1,1}^{f}(t)$ and $p_{2}(t) = S_{2,1}^{a}(t)*S_{1,2}^{a}(t)*S_{2,1}^{f}(t)*S_{1,2}^{f}(t)$ with * denoting the convolution operation. The total received signal at the RFID reader can be expressed as a summation of the coupling signal, *structural mode* and the *antenna mode* backscatters, as

$$y(t) = y_{c}(t) + y_{s}(t) + y_{a}(t) \tag{3.8}$$

Since the tag is at a distance R away from the reader, it causes the backscatter to arrive after propagation delay. The transmitted UWB pulse takes $\tau = \frac{R}{c}$ to reach to tag and also the resulting backscatter takes another equal time period to reach back to the reader. The total back-scatter $y_s(t) + y_a(t)$ can be expressed using the reflection coefficient presented by the tag to the incident UWB pulse transmitted by the reader unit. Therefore, Equation 3.8 can be expressed as in [9]

$$y(t) = y_c(t) + \alpha \Gamma_{in}(t) * x(t - \tau_1) \tag{3.9}$$

where $\tau_1 = \frac{2R}{c}$ is the propagation delay experienced by the transmitted pulse through its journey to and from the tag and α is the attenuation caused by the path loss. Substituting for $\Gamma_{in}(t)$ in Equation 3.9 using Equation 3.7, we get

$$y(t) = y_c(t) + \alpha S_{1,1}^a(t) * x(t - \tau_1) + \alpha p_1(t) * \\ x(t - \tau_1) + \alpha \Gamma_L p_2(t) * x(t - \tau_0 - \tau_1) \tag{3.10}$$

Since the second term, $\alpha S_{1,1}^a(t) * x(t - \tau_1)$, is multiplied by $S_{1,1}^a(t)$, it can be concluded that it is caused by the surface reflection of the incident wave illuminating the chipless tag as shown in the Figure 3.7. The third and fourth terms of Equation 3.10 are functions of both $S_{2,1}^a(t)$ and $S_{1,2}^a(t)$, which implies that these terms correspond to the signals that were har-nessed by the tag antenna (reason for having $S_{2,1}^a(t)$ terms) and that were in turn retransmitted by the tag antenna (reason for having $S_{1,2}^a(t)$ term). Therefore, we can conclude that the third and fourth terms of Equation 3.10 make up the *antenna mode* of the backscatter and that the second term makes up the *structural mode* backscatter. Hence, we can write

$$y_s(t) = \alpha S_{1,1}^a(t) * x(t - \tau_1) \quad \text{and} \\ y_a(t) = \alpha p_1(t) * x(t - \tau_1) + \alpha \Gamma_L p_2(t) * x(t - \tau_0 - \tau_1) \tag{3.11}$$

On further examination of the *antenna mode* backscatter expressed in (3.11), it is observed that only the second term of $y_a(t)$ contains $S_{2,1}^f(t)$ or $S_{1,2}^f(t)$. This implies that the second term contains the signal that has passed through the filter and was transformed by the filter transfer function $S_{2,1}^f(t)$. Therefore, it is clear that the useful information carried by the backscatter (the filter transfer function $S_{2,1}^f(t)$) is contained in the second term of the *antenna mode* backscatter. Also, we can see that the first and second terms of the *antenna mode* backscatter are sepa-rated by the delay, τ_0, introduced by the meandered transmission line.

Hence, the information-carrying component will be clearly separated and will be easily distinguishable from the other constituents of the backscatter.

3.4.2 Chipless RFID Tag Design

All the passive planar microwave circuits were designed for Taconic TLX0, a substrate material having a dielectric constant (εr) of 2.45 and loss tangent ($\tan\delta$) of 0.0019. The substrate thickness is 0.5 mm and copper cladding thickness is 17 μm. A coplanar waveguide (CPW) design approach was used for the chipless tag, which demands only a single surface of copper, where both the ground and other resonant structures lie in the same plane.

A coplanar circular disc-loaded monopole antenna, shown in Figure 3.8a, was designed and used as the tag antenna and transmit and

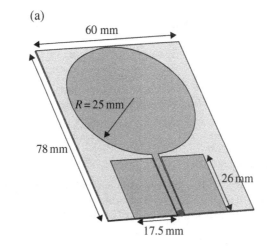

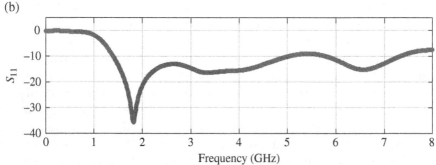

Figure 3.8 (a) Broadband antenna dimensions [5]. (b) Return loss profile.

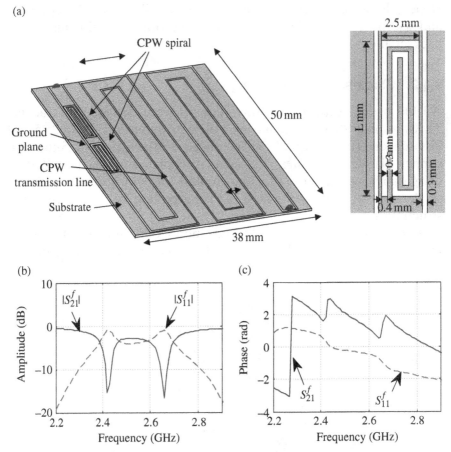

Figure 3.9 (a) Structure of the CPW spiral filters and meandering transmission line. (b) CPW spiral filter design. (c) CST simulation results of the amplitude and phase spectrums of the multiresonator filter transfer function [5].

received antennas at the RFID reader. The antenna return loss is shown in Figure 3.8b where it shows good performance from 1.4 to 7 GHz with a return loss of 10 dB or more.

A set of planar CPW spiral resonators were used to construct the multiresonator-based microwave filter that encodes information onto the signals captured by the tag antenna. Figure 3.9 shows the spiral resonator design, where the CPW spiral resonators are placed inside the transmission line. Figure 3.9a shows placement of the multiresonator filter along with the meandered transmission line. The filter consists of two spiral resonators one resonating at 2.42 GHz and the other resonating at 2.66 GHz. The simulation results of the filter are shown in Figure 3.9c where the two resonances are visible as sharp attenuations in the amplitude spectrum and as abrupt phase jumps in the phase spectrum of the filter transfer function $\left(S_{2,1}^{f}\right)$. The meandering transmission line has a length of 30.4 cm from the point where it connects to the monopole antenna. Therefore, the signals captured by the tag antenna will effectively travel double this distance before being retransmitted as the *antenna mode* backscatter. The propagation delay due to traveling this distance along the transmission line is approximately 3.2 ns.

3.5 SIMULATION RESULTS

In order to obtain a simulation result that is as close as to realistic conditions, the total system shown in Figure 3.6 was constructed as a 3D model in CST Microwave Studio™, and full-wave electromagnetic simulations were carried out. The simulation setup constructed in CST is shown in Figure 3.10 where the chipless RFID tag is placed 45 cm away from the reader antennas, which are themselves separated by 6 cm from each other.

The simulated total received signal is shown in Figure 3.11. Three cases were considered in the simulation where there is no tag present in front of the reader (shown in dark solid), chipless tag present in front of

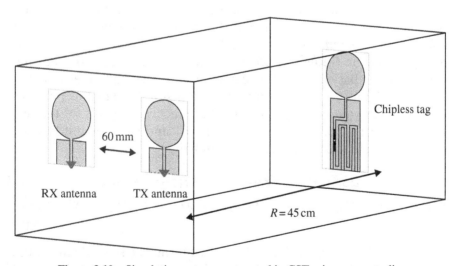

Figure 3.10   Simulation setup constructed in CST microwave studio.

the reader and terminated with an open circuit (shown in light solid), and chipless tag present in front of the reader with a short-circuited termination (shown in dark dashed). In all these three cases, the first and strongest component of the received signal is the coupled signal, $y_c(t)$. The backscatter is only present for the two cases where a tag is in front of the reader. The backscatter resulting from the presence of the chipless RFID is enlarged and shown in Figure 3.11b. It is observed that for the two cases in the presence of the tag with the two different loading conditions, the first component of the backscattered signal remains the same and is unaffected by the change of the load. This observation confirms that this first component is the *structural mode* backscatter, which is invariant to the load terminating the antenna. However, the second component following the *structural mode* shows a 180° phase difference for the two different loading conditions. This observation clearly reinforces the effect of $\Gamma_L = \pm 1$ in Equation 3.11 and confirms that this second component is in fact the *antenna mode* backscatter that is dependent upon the load connected to the tag antenna. Also, a clear time delay is observed that separates the *structural mode* and *antenna mode* backscatters due to the presence of the meandering transmission line.

3.6   PROCESSING RESULTS

From Figure 3.11, it is clearly seen that each of the components forming the total received signal can be separated out. To extract the information contained in the received signal, the component that contains the information (*antenna mode* backscatter) needs to be windowed out and processed in the frequency domain. But first, let us confirm that the *antenna mode* backscatter indeed contains the information encoded in the tag in the form of the filter transfer function $S_{2,1}^f$. For this purpose, let us define $y_{oc}(t)$ and $y_{sc}(t)$ to be the total received signals for the two states where the load is open circuit $\Gamma_L = 1$ and short circuit $\Gamma_L = -1$, respectively. Hence, using (3.10), we can write

$$y_{oc}(t) = y_c(t) + \alpha S_{1,1}^a(t) * x(t - \tau_1) + \alpha p_1(t) * x(t - \tau_1) + \alpha p_2(t) * x(t - \tau_0 - \tau_1)$$
$$y_{sc}(t) = y_c(t) + \alpha S_{1,1}^a(t) * x(t - \tau_1) + \alpha p_1(t) * x(t - \tau_1) - \alpha p_2(t) * x(t - \tau_0 - \tau_1)$$
$$\tag{3.12}$$

From (3.12), it is clear that only the last term (the second part of the *antenna mode* backscatter) changes sign, and the rest of the signal remains the same. Let $u(t)$ be the difference between $y_{oc}(t)$ and $y_{sc}(t)$

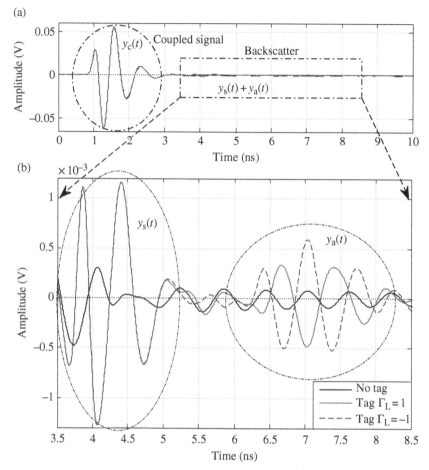

Figure 3.11 (a) The total received signal at the reader. (b) Enlarged section of the received signal showing the backscatter resulting due to the presence of the chipless RFID tag [5].

where all the other unwanted signals cancels out, and only the portion that carries the information $S_{2,1}^f$ remains:

$$u(t) = y_{oc}(t) - y_{sc}(t) = 2\alpha S_{2,1}^a(t) * S_{1,2}^a(t) * S_{2,1}^f(t) * S_{1,2}^f(t) * x(t - \tau_0 - \tau_1)$$

$$(3.13)$$

Figure 3.12a and c shows the spectral content of $u(t)$ obtained through simulation results shown in Figure 3.11. Here, the resonances at 2.42 and 2.66 GHz are clearly visible as two sharp attenuations in the amplitude spectrum and abrupt phase changes in the phase spectrum.

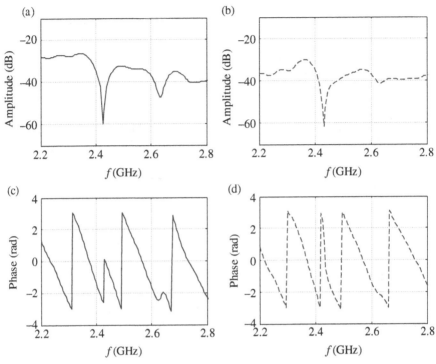

Figure 3.12 Tag frequency signature information extracted from the backscattered signal. The spectral content of $u(t)$ is shown in (a) and (c). The spectral content of a windowed portion of the *antenna mode* (with the coupled signal removed) is shown in (b) and (d) [5].

This confirms that the information is contained in the *antenna mode* backscatter.

In the process of estimating the tag signature from the received signal, the first step is to remove the residual effects of the transients caused by the coupled signal. In Figure 3.11b, when the three signals are compared, it is observed that the transients of the coupled signal (shown in dark solid) is comparable with the other two signals for which the tag is present (light-dashed and dark-dashed). In order to remove interference and improve SNR, this coupled signal, which is obtained through a tagless situation, needs to be subtracted from the total received signal, which can be either $y_{oc}(t)$ or $y_{sc}(t)$ depending on the termination condition of the chipless tag. Afterwards, the *antenna mode* backscatter, which starts at around the mark of 5.5 ns in Figure 3.11b, is windowed out using a raised cosine window, and its spectral content is obtained by applying the DFT using the fast Fourier Transform (FFT) algorithm. Figure 3.12b and d shows the spectral content obtained from such a windowed portion of the signal where both the amplitude and phase spectrums clearly show the presence of the two resonances that encodes information in the chipless RFID tag.

3.7 ANALYSIS OF BACKSCATTER FROM A MULTIPATCH-BASED CHIPLESS TAG

In this section, the backscatter from a multipatch-based chipless RFID tag is analyzed in the time domain [6]. A description of the overall system is provided. Next, the operation of the system and the design of the chipless RFID tag and the nature of the backscatter are explained using CST simulation results. Measurement results are also presented that validates the CST simulations. The performance of the system is evaluated for different tag positions with respect to the reader.

3.7.1 System Model and Expressions for Backscatter

The chipless RFID system considered is shown in Figure 3.13. The RFID reader consists of a single antenna that is used for both transmitting the interrogation pulse and for receiving the backscatter from the tag. The chipless RFID tag consists of a set of N inset-fed square patch antennas. Each patch antenna resonates at a unique frequency

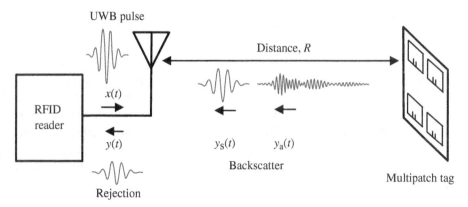

Figure 3.13 System model of the RFID system for reading multipatch-based chipless RFID [6].

$f_i, i = 1,...,N$. Depending on the combination of patches contained in the chipless tag, it is possible to encode N bits of information. Let $x(t)$ be the transmitted UWB pulse. When $x(t)$ is injected into the transmit antenna, part of it is radiated into the wireless medium and another part of it, $y_r(t)$, is rejected by the antenna, which is defined by the return loss profile or the $S_{1,1}(f)$ parameter of the antenna. Using the definition of $S_{1,1}(f)$, we can write

$$S_{1,1}(f) = \frac{F\{y_r(t)\}}{F\{x(t)\}} \tag{3.14}$$

where $F\{\cdot\}$ is the FT operator. This rejected signal is the largest and first signal to appear at the antenna terminals, which gradually decays down to zero. By this time, the radiation that was emitted by the antenna would have reached the tag where it interacts with the tag to generate backscatter. When this backscatter reaches back to the reader antenna, the transients of the rejection, $y_r(t)$, would have decayed and the antenna would be receptive to the backscatter incident on it. Therefore, the total received signal at the reader antenna can be expressed as

$$y(t) = F^{-1}\{S_{1,1}(f)X(f)\} + y_s(t) + y_a(t) \tag{3.15}$$

where $X(f) = F\{x(t)\}$. As explained in the previous sections, the structural mode backscatter, $y_s(t)$, arrives first, which is closely followed by the antenna mode backscatter, $y_a(t)$. Equation 3.15 can also be defined as follows:

$$y(t) = F^{-1}\{S_{1,1}^{L}(f)X(f)\} \tag{3.16}$$

where $S_{1,1}^{L}(f)$ is the modified return loss profile of the reader antenna due to the presence of the chipless tag in front of it. Here, we consider the backscatter produced by the chipless tag, $y_s(t) + y_a(t)$, to also be a part of the signal rejected by the antenna, where the tag has electromagnetically loaded or affected the antenna performance causing more rejection than in its original unaffected state. Using Equations 3.15 and 3.16, we can express the backscatter produced by the chipless tag as

$$y_s(t) + y_a(t) = F^{-1}\{[S_{1,1}^{L}(f) - S_{1,1}(f)]X(f)\} \tag{3.17}$$

3.7.2 The Design and Operation of the Multipatch-Based Chipless RFID

For demonstrating the operation of the system, a 4-bit chipless RFID tag consisting of four resonating patch antennas was considered. Each patch resonates at a unique frequency that can be controlled by varying the length L of the patch. Depending on the patches present in the tag, it can be engineered to have a unique spectral signature. With N distinct patches, it is possible to produce 2^N different unique spectral signatures enabling the encoding of N bits of information in each tag. Figure 3.14 shows the design of the multipatch-based chipless RFID tag. It is designed on a TLX8 substrate material having a dielectric constant of 2.55, loss tangent 0.0019, thickness of 0.5 mm, and copper cladding thickness of 17 μm. The tag contains four inset-fed square microstrip patch antennas. The four patch antennas of the

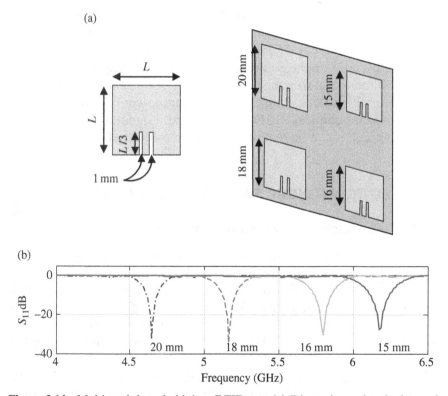

Figure 3.14 Multipatch-based chipless RFID tag. (a) Dimensions of a single patch and organization of the chipless tag. (b) Resonance frequencies of individual patch antennas [6].

tag have lengths 20, 18, 16, and 15 mm and resonate at 4.64, 5.16, 5.8, and 6.2 GHz, respectively.

When the transmitted UWB pulse reaches the tag, it gives rise to a *structural mode* backscatter, $y_s(t)$, due to the surface reflections occurring at the tag. A portion of the energy contained in the UWB pulse is captured by each of the patch antennas contained in the chipless tag. The tag antennas are terminated with an open-circuit load condition in order to maximize the mismatch at their termination. Hence, these captured signals will be eventually retransmitted by the same antennas to form the *antenna mode* backscatter $y_a(t)$. The *antenna mode* backscatter produced by the tag can be expressed as

$$y_a(t) = \sum_{i=1}^{N} y_a^i(t) \tag{3.18}$$

where $y_a^i(t)$ is the *antenna mode* backscatter generated by the i-th patch antenna of the tag. Each patch antenna resonates and receives signals at a unique frequency f_i. Therefore, the *antenna mode* backscatter produced by the i-th patch antenna will predominantly contain energy corresponding to the frequency f_i. Hence, by controlling the absence or presence of the patches and creating a unique combination of patch antennas for the chipless RFID tag, the spectral signature of the *antenna mode* backscatter can be readily engineered as desired.

3.8 ELECTROMAGNETIC SIMULATION OF SYSTEM

3.8.1 Four-Patch Backscattering Chipless Tag

Figure 3.15a shows the simulation setup used for simulating the RFID system using CST Microwave Studio™. The circular disc-loaded monopole antenna shown in Figure 3.8 is used at the reader as the transmit/receive antenna. The tag is placed 30 cm in front of the reader antenna. A Gaussian-modulated pulse having a bandwidth of 6 GHz (having frequency content from 2 to 8 GHz) was used for interrogating the chipless tag (see Fig. 3.15b).

The total received signal at the reader antenna is shown in Figure 3.16. The first and largest component is the rejected signal by the transmit antenna, $y_r(t)$, which gradually decays and fades away. At the 2.5 ns mark, a small echo is observed; this is the backscatter coming back from

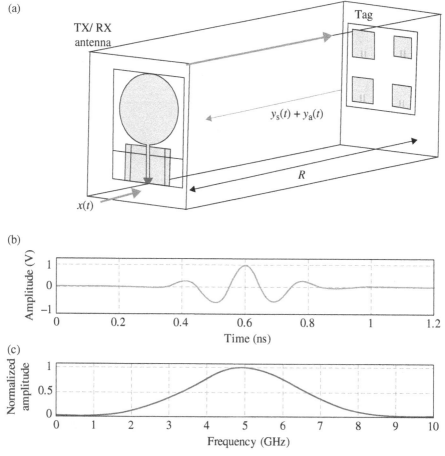

Figure 3.15 (a) Simulation setup used for simulating the multipatch-based chipless RFID system. (b) Time-domain view of UWB Interrogation pulse. (c) Amplitude spectrum of interrogation pulse [6].

the chipless tag. Compared to the rejected signal, the backscatter is very small and cannot be clearly seen in Figure 3.16a. An enlarged version of the backscatter is shown in Figure 3.16b. The backscatter consists of a large short-duration initial component followed by a smaller longer-duration component. The larger component has a shape similar to the transmitted Gaussian pulse shown in Figure 3.15b. We hypothesize that this larger component is the *structural mode* of the backscatter and that the smaller component is the *antenna mode*. However, because there is no transmission line introducing a controlled delay, as in the chipless RFID tag described in the previous section, it is not possible to clearly

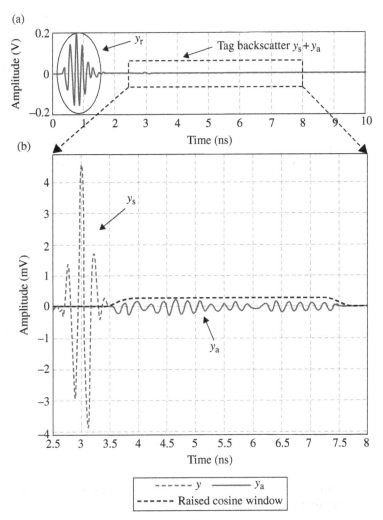

Figure 3.16 (a) Total received signal. (b) Enlarged portion of the backscatter showing the *structural mode* and *antenna mode* backscatter [6].

define where the *structural mode* backscatter ends and where the *antenna mode* backscatter starts. In order to confirm this hypothesis, these components need to be isolated, and their spectral content needs to be examined.

The larger (*structural mode* backscatter) and the smaller (*antenna mode* backscatter) components of the received backscatter were separated approximately using a raised cosine window as shown in Figure 3.16b. The DFT of these separated time-domain signals were then computed

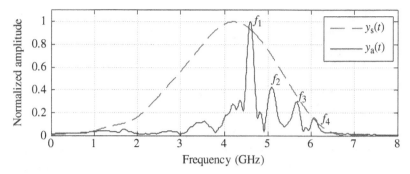

Figure 3.17 Amplitude spectra of the windowed *structural mode* and *antenna mode* backscatter obtained using FFT [6].

using the FFT algorithm. Figure 3.17 shows the spectral content of these windowed time-domain backscatter. The amplitude spectrum of the larger initial backscatter (Fig. 3.17, light-dashed line) has a Gaussian shape similar to the amplitude spectrum of the transmitted pulse shown in Figure 3.15c. This confirms the fact that the initial larger component is the *structural mode* backscatter. This is due to the fact that it is simply a reflection of the transmitted pulse having a similar spectral composition to the transmitted pulse without providing any information about the resonant properties of the patch antennas contained in the chipless tag. On the other hand, the amplitude spectrum of the secondary smaller component of the backscatter (shown in solid darker line in Fig. 3.17) following the *structural mode* clearly shows four spectral peaks exactly at the resonant frequencies of the patch antennas contained in the chipless tag. Therefore, it is evident that the secondary smaller- and longer-duration component of the backscatter is indeed the *antenna mode*, which is formed by the retransmitted energy from the patch antennas.

Also, it is observed that the contour formed by joining the spectral peaks of the *antenna mode* closely follow the contour of the Gaussian amplitude spectrum of the *structural mode* backscatter. The *antenna mode* backscatter is essentially a filtered version of the UWB pulse incident on the tag, where only the frequencies corresponding to the patches will be passed through while the others will be filtered out. Therefore, the heights of the spectral peaks of the *antenna mode* backscatter corresponding to frequencies having more energy (f_1) in the incident UWB pulse will be higher than those peaks corresponding to frequencies having lesser energy (f_2) in the incident UWB pulse. Here, the gain of the patch antennas also plays a role. If the gain is significantly higher for f_2 than f_1, this observation seen in Figure 3.17 might be different. Using a dynamic windowing technique with frequency reconfigurability, we can achieve uniform response for all frequency signatures.

3.8.2   Investigation into Reading Distance and Orientation of Tag

Figure 3.18 shows the frequency spectrum of the *antenna mode* backscatter from simulation results obtain for different positions and orientations of the tag with respect to the reader. The tag was placed in front of the reader antenna with four degrees of freedom as shown in

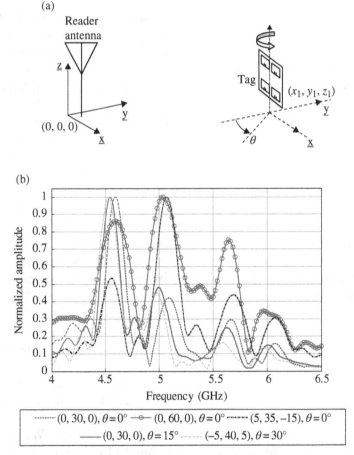

Figure 3.18   (a) Position of the tag with respect to the RFID reader. (b) Normalized amplitude spectrum of *antenna mode* backscatter for different tag locations (all distances are measured in centimeters and angles are in degrees) [6].

Figure 3.18a (x, y, z directions and rotation about z axis). All the four resonant spectral peaks can be clearly seen for all the different tag positions. However, a slight frequency shift is observed in the spectral peaks, particularly in the noncentered (tag locations not having $[0, y, 0]$ type coordinates) and rotated tag placements. The effect of this shift on the performance of decoding information contained in the spectral signature can be reduced by using a sufficiently large enough guard band between the resonances (100 MHz).

3.8.3 Measurement Results

Measurements were obtained for the multipatch-based chipless RFID using a vector network analyzer in an anechoic chamber environment as shown in Figure 3.19. Even though the vector network analyzer transmits signals and samples received signals in the frequency domain through the use of signal processing, this frequency-domain measurement information can be transformed to the time domain for our analysis as explained in Section 3.4.

The process of measurement involves two steps. First, the unaffected return loss profile, $S_{1,1}(f)$, of the reader antenna needs to be measured in a tagless environment. Afterwards, the tag is placed and the return loss of the electromagnetically loaded antenna, $S_{1,1}^{L}(f)$, is measured. No calibration tag measurements are required in this process. Using Equation 3.17, the time-domain total backscattered signal, $y_s(t) + y_a(t)$,

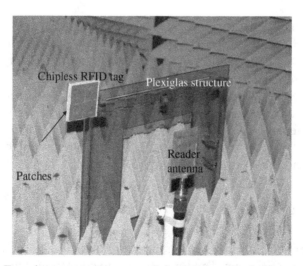

Figure 3.19 Experiment setup in an anechoic chamber. Tag is 30 cm away from the reader antenna.

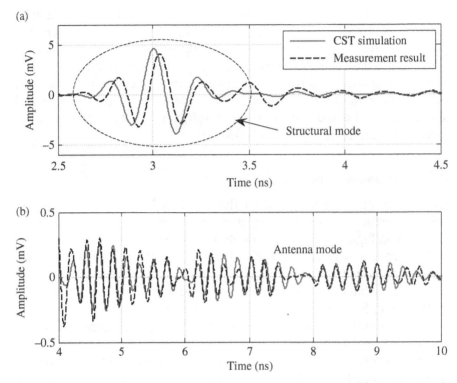

Figure 3.20 Time-domain (a) *structural mode* backscatter and (b) *antenna mode* backscatter obtained through simulation and the DFT of the frequency-domain measurement data.

can be calculated from these measurement data, which is shown in Figure 3.20. From this time-domain information, the *antenna mode* and *structural mode* backscatters can be windowed out in order to examine their frequency spectra.

Figure 3.21 shows the amplitude spectrums of the *structural mode* backscatter and the *antenna mode* backscatter that were calculated from the measured data. The simulated results are also plotted for comparison. We can see that there is a close agreement between the measured and the simulation results.

3.9 CONCLUSION

This chapter presents a time-domain view in analyzing the backscatter from chipless RFID tags. The limitations and challenges faced in the use of frequency-domain continuous wave readers can be overcome by

(a)

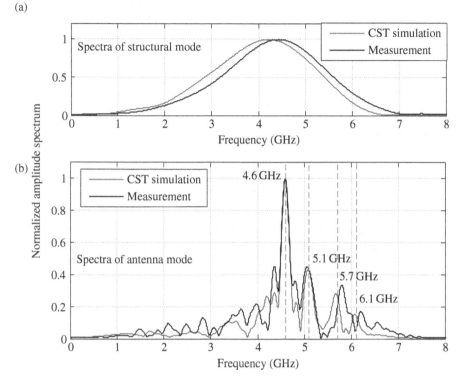

Figure 3.21 Amplitude spectrums of the (a) *structural mode* backscatter and (b) *antenna mode* backscatter obtained through measurement data and simulation data.

converting the frequency domain information into the time domain. This allows the isolated processing of the essential information-carrying part of the backscattered signal avoiding the interference caused by the noninformation-carrying components. UWB-IR-based readers also enable the reading of frequency signature-based chipless RFID tags where the received signal is directly sampled in the time domain using a high-speed analog to digital converter. Through the use of time-domain analysis, it is expected to improve the range and accuracy of the detection process involved at the RFID reader. The concepts involved in the time-domain analysis are demonstrated using two types of chipless RFID systems. The first type uses a multiresonator-based filter in order to encode information into the backscattered signal. The second type uses the frequency selective nature of the tag antennas in order to encode information into the backscattered signal. Using full-wave electromagnetic simulations carried out using CST microwave studio and measurements obtained from experiments conducted using a vector network analyzer, it was shown that the information encoded by the chipless tag is contained in the *antenna mode* of the backscatter. The time domain-based processing of the backscatter allows the extraction of information contained in the chipless RFID tag without the need of calibration tags.

REFERENCES

1. S. Preradovic and N. C. Karmakar, "*Design of short range chipless RFID reader prototype*," presented at the 5th International Conference on Intelligent Sensors, Sensor Networks and Information Processing (ISSNIP), 2009, Melbourne, Australia, December 7–10, 2009.
2. S. Preradovic, *et al.*, "*A novel chipless RFID system based on planar multi-resonators for barcode replacement*," presented at the IEEE International Conference on RFID, 2008, Las Vegas, Nevada, 2008.
3. C. A. Balanis, *Antenna Theory Analysis and Design*, 3rd edition. Hoboken, NJ: John Wiley & Sons, Inc., 2005.
4. I. Balbin and N. C. Karmakar, "Phase-encoded chipless RFID transponder for large-scale low-cost applications," *IEEE Microwave and Wireless Component Letters*, vol. 19, pp. 509–511, 2009.
5. P. Kalansuriya and N. C. Karmakar, "*Time domain analysis of a backscattering frequency signature based chipless RFID tag*," presented at the IEEE Asia-Pacific Microwave Conference Proceedings (APMC), 2011, Melbourne, Australia, December 5–8, 2011.
6. P. Kalansuriya and N. C. Karmakar, "*UWB-IR based detection for frequency-spectra based chipless RFID*," presented at the IEEE MTT-S International Microwave Symposium Digest (MTT), 2012, Montreal, Canada, June 17–22, 2012.
7. I. Balbin and N. C. Karmakar, "Multi-Antenna Backscattered Chipless RFID Design," in *Handbook of Smart Antennas for RFID Systems*, N. C. Karmakar, Ed., Hoboken, NJ: John Wiley & Sons, Inc., 2010, pp. 413–443.
8. S. Hu, C. L. Law, and W. Dou, "A balloon-shaped monopole antenna for passive UWB-RFID tag applications," *IEEE Antennas and Wireless Propagation Letters*, vol. 7, pp. 366–368, 2008.
9. A. Ramos, *et al.*, "Time-domain measurement of time-coded UWB chipless RFID tags," *Progress in Electromagnetics Research*, vol. 116, pp. 313–331, 2011.
10. S. Hu, Y. Zhou, C. L. Law, and W. Dou, "Study of a Uniplanar Monopole Antenna for Passive Chipless UWB-RFID Localization System," *IEEE Transactions on Antennas Propagation*, vol. 58, pp. 271–278, 2010.

CHAPTER 4

SINGULARITY EXPANSION METHOD FOR DATA EXTRACTION FOR CHIPLESS RFID

4.1 INTRODUCTION

This chapter introduces the singularity expansion method (SEM) [1] and its application in chipless RFID. The SEM was introduced in order to explain some early experimental observations in experiments involving high energy electromagnetic pulses (EMPs). The approach expresses the scattered signals from objects that were subjected to EMPs as a sum of damped sinusoidal signals. This observation is strikingly similar to the electromagnetic interaction that occurs when a chipless RFID tag is being interrogated using an ultra-wide bandwidth impulse of electromagnetic energy. Therefore, SEM is directly applicable in explaining the behavior of chipless RFID.

The use of SEM for chipless RFID has been explored recently in the work of Refs. [2, 3] where the backscattered signal from a chipless RFID tag is characterized using a set of pole singularities and the accompanying complex amplitudes or residues. This representation provides more dimensions or degrees of freedom in encoding information. The extraction of poles and residues from a sampled time-domain signal involves some mathematical processing of the signal. It is a well-studied

Chipless Radio Frequency Identification Reader Signal Processing, First Edition.
Nemai Chandra Karmakar, Prasanna Kalansuriya, Rubayet E. Azim and Randika Koswatta.
© 2016 John Wiley & Sons, Inc. Published 2016 by John Wiley & Sons, Inc.

topic and several efficient methods exist that enables the accurate estimation of the poles signature of a time-domain signal. The matrix pencil algorithm is widely used for this purpose and it is based on the discrete time Laplace transformation.

The rest of the chapter is organized as follows. Section 4.2 provides an introduction of SEM and the representation of damped sinusoids in the complex frequency domain. The extraction of poles and residues from a sampled time-domain signal is explained in detail in Section 4.3 where the operation of the matrix pencil algorithm is discussed. Section 4.4 discusses the application of SEM through the matrix pencil algorithm in chipless RFID based on the recent work of Refs. [2, 3]. Finally, conclusions are drawn in Section 4.5 (Fig. 4.1).

4.2 THE SEM

The SEM was introduced by Carl Baum in 1971 [1] as an approach to represent the solution of an electromagnetic interaction using a set of singularities in the complex frequency plane or the s-domain associated with the Laplace transform. In particular, it can be used in the characterization or calculation of the transient responses of antennas or other passive scatterers of electromagnetic radiation, which makes it a directly applicable tool in the analysis of backscatter from chipless RFID. SEM represents the analytic properties of an electromagnetic response through the double-sided Laplace transform. The singularities of the complex frequency domain obtained through the Laplace transform are used in the characterization of the response, hence the name SEM.

The early development of the SEM [4, 5] was based on experimental observations of objects that were subjected EMP experiments, where they were exposed to a broadband transient electromagnetic excitation composed of a continuum of frequencies. The observed late-time transient response waveforms appear to be dominated by a few damped sinusoids [6]. A damped sinusoid corresponds to a pair of pole singularities in the complex frequency plane, which implies that the object that scatters the energy has a large response or resonance at frequencies near these poles. It was also observed that these pole singularities associated with a scattering object are independent of the incident source field and observer location, where they are inherent properties of the source-free natural frequencies of the object. This makes the pole singularities associated with a scatterer useful for its remote identification [6].

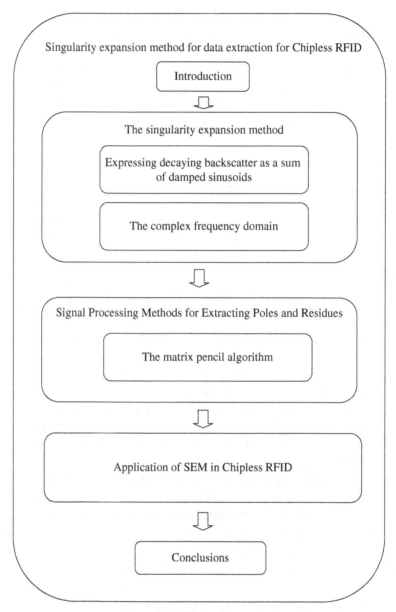

Figure 4.1 Chapter overview.

From the time-domain analysis detailed in the previous chapter, it is clear that the backscatter produced by a scatterer, when illuminated with electromagnetic radiation, consists of two parts, namely, the *structural mode* and the *antenna mode* backscatter or radar cross section

(RCS). From the analysis presented in Chapter 3, it is evident that the structural mode backscatter arrives earlier and its spectral content is determined by the excitation pulse, whereas the spectral content of the *antenna mode* backscatter that arrives at a later stage in time following the earlier echo is essentially determined by the resonant properties of the scatterer. In essence, the SEM simply expresses this late-time, source-independent backscatter, $y(t)$, as a sum of damped sinusoids as follows [6]:

$$y(t) = \sum_{i=1}^{P} a_i e^{-\sigma_i t} \cos(\omega_i t + \phi_i) \tag{4.1}$$

where a_i and ϕ_i are the source- and observation-dependent amplitude and phase, σ_i is the damping factor, ω_i is the angular frequency of the ith damped sinusoid, and P is the number of sinusoids. As opposed to a simple spectral analysis of the late-time response through Fourier analysis, the SEM approach goes a step further and also provides information on the rate of damping associated with the natural resonances or oscillations of a scatterer.

The principles of SEM are directly applicable in the domain of chipless RFID since a chipless RFID is also a passive scatter of electromagnetic energy that is illuminated with an electromagnetic pulse. It is clear from the discussion presented in Chapter 3 that the backscattered signals from a chipless RFID tags are of transitory and decaying nature. Therefore, they can be readily expressed using Equation 4.1. The definition of SEM is more suited for chipless RFID systems that use ultra-wideband impulse radar for interrogation where the tag is excited using an impulse of electromagnetic energy. However, even with systems that use frequency swept-based interrogation, if the reader is capable of accurately extracting both amplitude and phase of the received signal in the frequency domain, the equivalent time-domain response for any given impulse can be obtained using inverse Fourier analysis through signal processing. This response can then be used for SEM analysis.

4.2.1   The Complex Frequency Domain

A time-domain signal can be represented in the complex frequency domain through the Laplace transform. The double-sided Laplace transform of $x(t)$ is defined as follows:

$$X(s) = L\{x(t)\} = \int_{-\infty}^{\infty} x(t) e^{-st} dt \tag{4.2}$$

where $s = \sigma + j\omega$ is the complex frequency. The inverse Laplace transform is used to transform a function $X(s)$ in the complex frequency plane to the time domain:

$$x(t) = L^{-1}\{X(s)\} = \frac{1}{2\pi j} \lim_{\lambda \to \infty} \int_{\eta - j\lambda}^{\eta + j\lambda} X(s) e^{st} ds \qquad (4.3)$$

Figure 4.2 shows some time-domain signals and the corresponding representation in the complex frequency domain. Passive backscattered signals emanating from scatterers are usually decaying and transitory. Therefore, the poles corresponding to such signals have their poles located in the left side of the complex frequency plane with a negative real part. Signals having purely imaginary poles are lossless and do not experience any decaying with time. Unstable signals that are exponentially growing with time will have their poles located in the right side of the complex frequency plane.

Consider the sum of damped sinusoidal signals expressed in Equation 4.1. Let $Y(s)$ denote the Laplace transform of $y(t)$ where

$$Y(s) = L\{y(t)\} = \int_{-\infty}^{\infty} y(t) e^{-st} dt = \sum_{i=1}^{P} \frac{a_i e^{j\phi_i}}{2(s + \sigma_i - j\omega_i)} + \frac{a_i e^{-j\phi_i}}{2(s + \sigma_i + j\omega_i)} \qquad (4.4)$$

Each damped sinusoid gives rise to two pole singularities $-\sigma_i + j\omega_i$ and $-\sigma_i - j\omega_i$ in the complex frequency domain. The complex amplitudes $(a_i e^{j\phi_i})/2$ and $(a_i e^{-j\phi_i})/2$ that are associated with each of the poles are also called residues associated with each of the singularities. A more general version of Equation 4.1 can also be written as a weighted sum of complex exponentials as follows [7]:

$$y(t) = \sum_{i=1}^{M} R_i e^{s_i t} \qquad (4.5)$$

where $s_i = -\sigma_i + j\omega_i$ are the poles of the signal, R_i are the complex amplitudes or residues associated with each pole, and M is the number of poles. It should be noted that the poles s_i appear as complex conjugate pairs since $y(t)$ is a real signal. Let $y(t)$ be sampled every Δt, which gives the following sampled time-domain signal:

$$y(t_k) = y(k\Delta t) = \sum_{i=1}^{M} R_i e^{s_i(k\Delta t)}; \quad k = 0, 1, \ldots, N-1 \qquad (4.6)$$

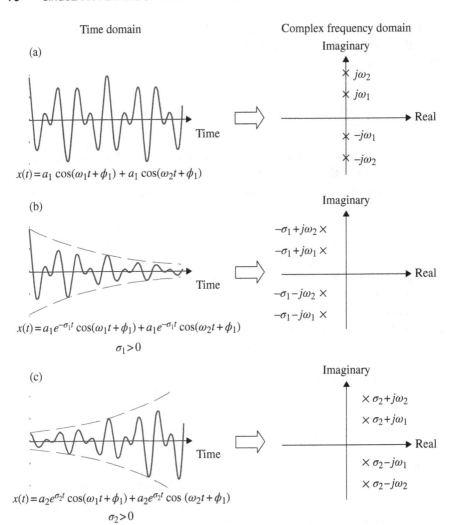

Figure 4.2 Time-domain and complex frequency domain representations of damped sinusoidal signals. (a) Undamped sinusoidal signal. (b) Damped decaying sinusoidal signal. (c) Unstable exponentially growing sinusoid.

Equation 4.6 can be rewritten as [7, 8]

$$y(t_k) = \sum_{i=1}^{M} R_i \, z_i^k; \quad k = 0, 1, \ldots, N-1 \tag{4.7}$$

where the parameters z_i are defined as $z_i = e^{s_i \Delta t} = e^{(-\sigma_i + j\omega_i)\Delta t}$.

4.2.2 Extraction of Poles and Residues

This section presents the mathematical tools that enable the extraction of the complex frequency domain poles from sampled time-domain data. The objective of such a mathematical algorithm is to find the best estimates of M, s_i, and R_i in Equation 4.7 from a given set of sampled data that may be contaminated with noise.

Two methods are reported in the literature in solving this problem: Prony's polynomial algorithm [9, 10] and the matrix pencil algorithm [7, 8, 11]. The Prony's polynomial algorithm employs a two-step approach, where first, the coefficients of a polynomial are calculated by solving a matrix equation consisting of the data samples. Next, the roots of this polynomial, which are equal to the z_i parameters, are found that are then used in calculating the poles s_i. This method was introduced by Prony in 1795 and was later further improved in Refs. [9, 10]. Prony's algorithm provides an efficient and accurate method to estimate the poles and residues from a set of sampled data, which are equally spaced in time. However, it is extremely sensitive to noise present in the data where modifications were introduced in Refs. [12–14] in order to improve its performance under noise. The matrix pencil method is a one-step approach and involves in solving a generalized eigenvalue problem in order to obtain the z_i parameters from the sampled data. Therefore, with the matrix pencil approach, there is no practical limitation on the number of poles that can be found. Next, we will present the operation of the matrix pencil algorithm and how it can be applied to extract the poles and residues from a time-domain signal.

4.2.3 Matrix Pencil Algorithm

First, let's define two matrices, $[Y_0]$ and $[Y_1]$, using the sampled time-domain signal in Equation 4.7 as [7, 8]

$$[Y_0] = \begin{bmatrix} y_0 & y_1 & \cdots & y_{L-1} \\ y_1 & y_2 & \cdots & y_L \\ \vdots & \vdots & \ddots & \vdots \\ y_{N-L-1} & y_{N-L} & \cdots & y_{N-2} \end{bmatrix}_{(N-L) \times L}$$

$$\text{and} \quad [Y_1] = \begin{bmatrix} y_1 & y_2 & \cdots & y_L \\ y_2 & y_3 & \cdots & y_{L+1} \\ \vdots & \vdots & \ddots & \vdots \\ y_{N-L} & y_{N-L+1} & \cdots & y_{N-1} \end{bmatrix}_{(N-L) \times L}$$

(4.8)

where y_k denotes the kth sample $y(t_k)$ in Equation 4.6 and L is the pencil parameter. This parameter plays an important role in eliminating the detrimental effects of noise contaminating the sampled data. The matrices $[Y_0]$ and $[Y_1]$ can be expressed as [7, 8]

$$[Y_1]=[Z_1][R][Z_0][Z_2] \quad \text{and} \quad [Y_0]=[Z_1][R][Z_2] \tag{4.9}$$

where

$$[Z_1]=\begin{bmatrix} 1 & 1 & \cdots & 1 \\ z_1 & z_2 & \cdots & z_M \\ \vdots & \vdots & \ddots & \vdots \\ z_1^{(N-L-1)} & z_2^{(N-L-1)} & \cdots & z_M^{(N-L-1)} \end{bmatrix}_{(N-L)\times M},$$

$$[Z_2]=\begin{bmatrix} 1 & z_1 & \cdots & z_1^{L-1} \\ 1 & z_2 & \cdots & z_2^{L-1} \\ \vdots & \vdots & \ddots & \vdots \\ 1 & z_M & \cdots & z_M^{L-1} \end{bmatrix}_{M\times L},$$

$$[Z_0]=\begin{bmatrix} z_1 & 0 & \cdots & 0 \\ 0 & z_2 & \cdots & 0 \\ \vdots & \vdots & \ddots & \vdots \\ 0 & 0 & \cdots & z_M \end{bmatrix}_{M\times M}, \text{and} [R]=\begin{bmatrix} R_1 & 0 & \cdots & 0 \\ 0 & R_2 & \cdots & 0 \\ \vdots & \vdots & \ddots & \vdots \\ 0 & 0 & \cdots & R_M \end{bmatrix}_{M\times M}$$

Using Equation 4.9, the linear matrix pencil $[Y_1]-\lambda[Y_0]$ can be expressed as [7, 8]

$$[Y_1]-\lambda[Y_0]=[Z_1][R]\{[Z_0]-\lambda[I]\}[Z_2] \tag{4.10}$$

where $[I]$ is the identity matrix of dimension $M \times M$. In general, it can be shown that the rank of the linear matrix pencil, $[Y_1]-\lambda[Y_0]$, will be M, provided that $M \leq L \leq N-M$ [11, 15, 16]. The rank of this matrix becomes $M-1$ if $\lambda = z_i$, $i = 1, 2, \ldots, M$. Therefore, it is clear that the parameters z_i can obtained by solving for the generalized eigenvalues of the matrix pair $\{[Y_1], [Y_0]\}$. This problem can also be formulated as solving for the ordinary eigenvalues of the matrix $[A]=[Y_0]^{\dagger}[Y_1]$ where $[Y_0]^{\dagger}$ is the Moore–Penrose pseudoinverse of $[Y_0]$, which is defined as follows:

$$[Y_0]^\dagger = \left\{[Y_0]^H[Y_0]\right\}^{-1}[Y_0]^H, \qquad (4.11)$$

with the superscript H denoting the operation of matrix conjugation. Once the z_is are calculated, the poles of the signal can be expressed as

$$s_i = \frac{\ln z_i}{\Delta t} \qquad (4.12)$$

Since the z_is and M are known by solving for the eigenvalues of $[Y_0]^\dagger[Y_1]$, the residues R_i can be calculated by solving the following matrix equation expressed using Equation 4.7:

$$\begin{bmatrix} y_0 \\ y_1 \\ \vdots \\ y_{N-1} \end{bmatrix} = \begin{bmatrix} 1 & 1 & \cdots & 1 \\ z_1 & z_2 & \cdots & z_M \\ \vdots & \vdots & \ddots & \vdots \\ z_1^{N-1} & z_2^{N-1} & \cdots & z_M^{N-1} \end{bmatrix} \begin{bmatrix} R_1 \\ R_2 \\ \vdots \\ R_M \end{bmatrix} = [Z] \begin{bmatrix} R_1 \\ R_2 \\ \vdots \\ R_M \end{bmatrix}$$

$$\begin{bmatrix} R_1 \\ R_2 \\ \vdots \\ R_M \end{bmatrix} = [Z]^\dagger \begin{bmatrix} y_0 \\ y_1 \\ \vdots \\ y_{N-1} \end{bmatrix} \qquad (4.13)$$

This technique of calculating the poles and residues provides good results only in the absence of noise and numerical errors in the sampled data. However, when the sampled data are contaminated by the slightest amount of noise or other sources of numerical error, the calculated singularities show a very large deviation from their accurate values. Therefore, to combat the adverse effect of numerical errors, the total least-squares matrix pencil (TLSMP) method [7] is used.

Here, the noise-contaminated matrices $[Y_0]$ and $[Y_1]$ are combined to produce the matrix $[Y]$ where $[Y_0]$ is obtained from $[Y]$ by deleting its last column and $[Y_1]$ is obtained from $[Y]$ by deleting its first column:

$$[Y] = \begin{bmatrix} y_0 & y_1 & \cdots & y_L \\ y_1 & y_2 & \cdots & y_{L+1} \\ \vdots & \vdots & \ddots & \vdots \\ y_{N-L-1} & y_{N-L} & \cdots & y_{N-1} \end{bmatrix}_{(N-L)\times(L+1)} \qquad (4.14)$$

For the efficient filtering of noise affecting the data, the pencil parameter L should be chosen to be between $N/3$ and $N/2$ [16–18]. Next, the singular value decomposition (SVD) of the matrix $[Y]$ is performed, which yields

$$[Y]=[U]\left[\sum\right][V] \tag{4.15}$$

where $[U]$ is an $(N-L)\times(N-L)$ unitary matrix consisting of the orthonormal eigenvectors of $[Y][Y]^H$ and $[V]$ is an $(L+1)\times(L+1)$ unitary matrix consisting of the eigenvectors $[Y]^H[Y]$. The matrix $[\Sigma]$ is a $(N-L)\times(L+1)$ diagonal matrix and contains the singular values of $[Y]$. If the data contained in the matrix $[Y]$ were noise-free, it would result in exactly M nonzero singular values. However, due to the presence of noise, the otherwise zero-valued singular values are perturbed to have small nonzero values. These small nonzero singular values result in errors in the poles calculated through the eigenvalues of $[Y_0]^{\dagger}[Y_1]$. Hence, in the TLSMP method in order to suppress the detrimental effects of noise, these spurious singular values are removed from the singular value matrix $[\Sigma]$ and also the corresponding eigenvectors are removed from the matrices $[U]$ and $[V]$. For an unknown signal contaminated with noise, of which the poles $s_i=-\sigma_i+j\omega_i$ need to be determined, the number of poles M of the original signal unaffected by noise is not known beforehand. Therefore, when removing the spurious singular values from $[\Sigma]$, the M' dominant singular values are chosen and the rest are discarded. The dominant singular values are chosen such that

$$\alpha_{max} > \alpha_i > 10^{-d}\ \alpha_{max} \tag{4.16}$$

where α_i is the ith singular value in $[\Sigma]$, α_{max} is the maximum singular value, and d is the significant decimal digits considered. Overestimation of M does not severely affect the solution, it would only produce additional poles in the complex frequency domain that have smaller magnitudes or residues, and the dominant poles will appear with larger residues. However, underestimation of M will cause large errors. Once a suitable estimate M' for M is determined, the new matrices $[\Sigma']$, $[U']$, and $[V']$ are obtained by removing the spurious singular values and the corresponding eigenvectors as follows [7]:

$$[U'] = [u_1 \quad u_2 \quad \cdots \quad u_{M'}]_{(N-L) \times M'} \quad [V'] = [v_1 \quad v_2 \quad \cdots \quad v_{M'}]_{(L+1) \times M'}$$

$$[\Sigma'] = \begin{bmatrix} \alpha_1 & 0 & \cdots & 0 \\ 0 & \alpha_2 & \cdots & 0 \\ \vdots & \vdots & \ddots & \vdots \\ 0 & 0 & \cdots & \alpha_{M'} \end{bmatrix}_{M' \times M'} \tag{4.17}$$

where u_i is the ith column vector in $[U]$ and v_i is the ith column vector in $[V]$. Using $[\Sigma']$, $[U']$, and $[V']$, the noise filtered versions of $[Y_0]$ and $[Y_1]$ can be expressed as [7]

$$[Y_0] = [U'][\Sigma'][V_0']^H \quad \text{and} \quad [Y_1] = [U'][\Sigma'][V_1']^H \tag{4.18}$$

where $[V_0']$ is the matrix obtained by removing the last column from $[V']$ and $[V_1']$ is the matrix obtained by removing the first column from $[V']$. The poles of the signal can now be obtained using the eigenvalues of the noise filtered $[Y_0]$ and $[Y_1]$ as before. Equivalently, they can also be obtained simply by solving for the ordinary eigenvalues of the matrix $B = \{[V_0']^H\}^\dagger [V_1']^H$. The nonzero eigenvalues of $\{[V_0']^H\}^\dagger [V_1']^H$ will be equal to the z_i parameters of the signal.

4.2.4 Case Study

Consider the following time-domain signal:

$$y_1(t) = e^{-\sigma_1 t} \cos(2\pi f_1 t + \phi_1) + e^{-\sigma_2 t} \cos(2\pi f_2 t + \phi_2); \quad t > 0 \tag{4.19}$$

with the following parameters—frequencies $f_1 = 2$ GHz and $f_2 = 3$ GHz, damping factors $\sigma_1 = 5 \times 10^8$ and $\sigma_2 = 1.5 \times 10^8$, and phases $\phi_1 = -\frac{\pi}{2}$ and $\phi_2 = -\frac{\pi}{2}$. Figure 4.3 shows the time-domain and complex frequency domain representation of this signal. The signal contains $M = 4$ pole singularities in the complex frequency plane that appear as two complex conjugate pairs.

The signal given in Equation 4.19 was contaminated with white Gaussian noise to give a signal-to-noise ratio (SNR) of −5 dB. This contaminated version of $y_1(t)$ was sampled every 2 ps to obtain $N = 1000$ noisy samples where the sampling started at $t_1 = 500$ ps. These noisy samples serve as the unknown signal of which the poles and residues need to be found. For the calculation, the pencil parameter L was

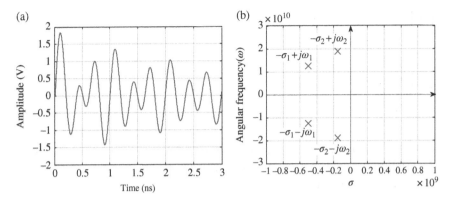

Figure 4.3 (a) Time-domain signal and (b) poles of the signal in the complex frequency plane.

Table 4.1 Singular Values of the Matrix [Y]

Index i	Singular Value α_i
1	200.09
2	197.54
3	119.52
4	110.47
5	45.16
6	45.1
7	43.54
8	43.47
9	43.16
10	43.05

The shaded region shows the dominant singular values.

chosen to be $N/2 = 500$. First, using these noisy samples, the matrix [Y] given in Equation 4.15 is formed and its singular values were obtained using SVD, which are shown in Table 4.1. It is clear that only the first four singular values are dominant. Therefore, we can easily estimate the number of poles contained in the signal to be four. However, for the sake of illustrating the methodology involved in the matrix pencil algorithm, let us overestimate M' to be six.

Using the process explained in Equations 4.17 and 4.18, the unitary matrix [V'] and subsequently the matrices [V_0'] and [V_1'] are obtained. By solving for the eigenvalues of $\{[V_0']^H\}^\dagger [V_1']^H$, the z_i parameters of the signal are obtained and in turn the poles s_i are calculated using Equation 4.12. The residues associated with the poles are computed using Equation 4.13.

Table 4.2   (a) Poles and Residues of the Original Signal and (b) Estimated Poles and Residues of the Noise-Contaminated Signal

Pole s_i	Residue R_i	Magnitude of Residue $|R_i|$
	(a)	
$-1.5 \times 10^8 + j1.88 \times 10^{10}$	$-j0.5$	0.5
$-1.5 \times 10^8 - j1.88 \times 10^{10}$	$j0.5$	0.5
$-5 \times 10^8 + j1.26 \times 10^{10}$	$-j0.5$	0.5
$-5 \times 10^8 - j1.26 \times 10^{10}$	$j0.5$	0.5
	(b)	
$-1.62 \times 10^8 + j3.99 \times 10^{10}$	$0.0543 - j0.046$	0.0712
$-1.62 \times 10^8 - j3.99 \times 10^{10}$	$0.0543 + j0.046$	0.0712
$-1.36 \times 10^8 + j1.89 \times 10^{10}$	$-0.0144 - j0.4662$	0.4664
$-1.36 \times 10^8 - j1.89 \times 10^{10}$	$-0.0144 + j0.4662$	0.4664
$-6.82 \times 10^8 + j1.26 \times 10^{10}$	$-0.0351 - j0.5839$	0.5849
$-6.82 \times 10^8 - j1.26 \times 10^{10}$	$-0.0351 + j0.5839$	0.5849

The shaded region shows the dominant poles having larger magnitude in their residues.

The estimated poles (s_i) and residues (R_i) of the noise-contaminated signal are given in Table 4.2b. The magnitudes of the residues corresponding to the first two poles are significantly smaller than the magnitudes of the residues of the other four poles. Therefore, it can be concluded that these first two poles are due to the noise contaminating the signal. From the results shown in Table 4.2, we can see a very close agreement between the original and estimated poles where there is only an error in the estimated damping factor.

Using these estimated poles and residues, the noisy signal can be reconstructed using Equation 4.6. Figure 4.4a shows the reconstructed signal and the noise-contaminated signal, and Figure 4.4b shows the reconstructed signal and the original signal $y_1(t)$. It is evident from the results shown in the figure that the reconstructed signal, estimated using the noise-contaminated data samples, matches almost perfectly with the original signal $y_1(t)$.

The original poles and the estimated dominant poles for different SNR values are shown in Figure 4.5. From the figure, it is observed that as the noise level affecting the signal increases, the estimated pole starts to deviate from its correct location. It is also observed that the deviation seen in pole that has a higher damping is greater than the pole that has a lower damping. This can be intuitively explained since the sinusoid having a higher damping coefficient will quickly decay and will be more

(a)

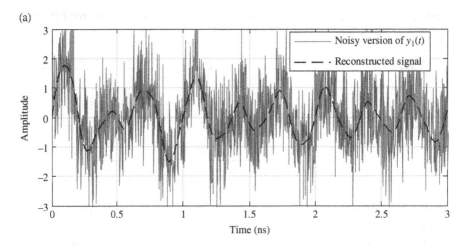

(b)

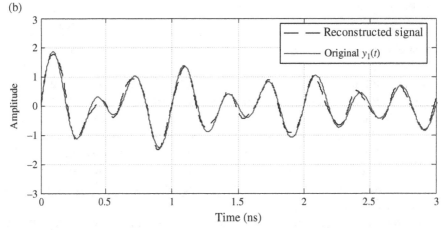

Figure 4.4 (a) Noise-contaminated signal (SNR = −5 dB) and reconstructed signal. (b) Original signal and reconstructed signal.

susceptible to perturbations caused by noise than a sinusoid having a lower damping coefficient. It is also observed that for all the estimations, the error in the angular frequency is negligible.

4.3   APPLICATION OF SEM FOR CHIPLESS RFID

This section discusses the use of SEM in chipless RFID. We will detail some recent developments reported in the literature [2, 3] on the use of the matrix pencil method for the representation of tag backscatter as a sum of damped sinusoids.

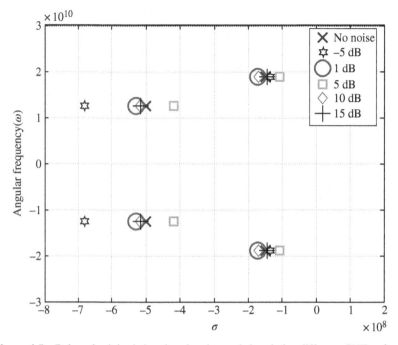

Figure 4.5   Poles of original signal and estimated signals for different SNR values.

In Refs. [2, 3], the authors make use of a notched elliptical dipole tag to demonstrate the use of SEM in encoding data into the tag and decoding it during the detection using the matrix pencil algorithm. Figure 4.6 shows the chipless notched elliptical dipole tag used in Ref. [2, 3]. The same set of notches is symmetrically repeated on the four corners of the elliptical dipole tag. The notches cause resonances to occur resulting in nulls in the frequency spectra of the backscattered response from the tag. It is shown in Ref. [19] that the occurrence of these resonant nulls can be readily controlled by the presence or absence of the physical notches etched out on the elliptical dipole. Therefore, each of these nulls can be used to encode a bit of information as in conventional frequency-domain chipless RFID tags. As oppose to simply viewing the tag response in the frequency domain, the authors of Refs. [2, 3] characterize the properties of these resonant nulls in the complex frequency domain as pole singularities using the matrix pencil algorithm. Therefore, each resonant null is defined as a complex pole having both real (damping coefficient) and imaginary (frequency) components. This gives rise to an additional dimension (damping coefficient) that can be used to encode more information or implement error correction mechanisms at the reader.

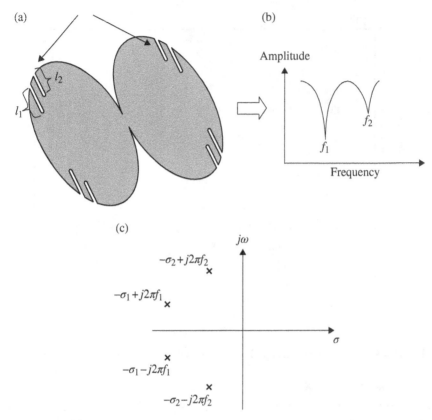

Figure 4.6 (a) Notched elliptical dipole chipless RFID tag having two notches giving rise to two resonances, (b) frequency-domain representation of tag response, and (c) representation of tag response as pole singularities in complex frequency plane.

In the system presented in Refs. [2, 3], the chipless tag is interrogated using a broadband swept frequency interrogation signal produced by a vector network analyzer (VNA). The tag response is given by the measured reverse scattering parameter $S_{12}(f)$. This frequency-domain measurement is first converted to the time domain. In order to obtain the pole signature of the tag from this time-domain backscatter signal, the late-time response that is free from the influence of the source excitation needs to be isolated. For this, the time-domain signal is windowed at a suitable position determined experimentally [3]. The matrix pencil algorithm is then applied to this windowed portion of the time-domain backscatter to extract the pole singularities.

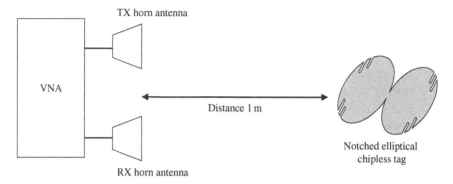

Figure 4.7 Experimental setup used for taking measurements from the notched elliptical chipless RFID tag in Ref. 3.

The measurement setup used for the experiments conducted in Ref. [3] is illustrated in Figure 4.7. The chipless RFID tag reader is essentially a VNA with two horn antennas attached to its two ports — one serving as the transmit antenna and the other serving as the receiving antenna. The horn antennas used in the experiment provide wide bandwidth (2–20 GHz and 1–18 GHz) and the antenna gain also varies considerably within this frequency range (5–19 dB and 6–14 dB). The tag was placed 1 m from the reader antennas and a transmission power of 20 dBm was used to interrogate the chipless tag.

Before the pole signature of the tag can be extracted from the measured $S_{12}(f)$, some processing is required where unwanted signals such as the coupling between the antennas and the background reflections need to be removed. For this purpose, the reading of the tag involves a three-step measurement process as follows:

i. The first measurement is done with an empty measurement (anechoic) chamber without the tag where the effect of the background reflections together with the antenna coupling is captured. Let $S_{12}(f)^{(1)}$ denote the first measurement taken, it can be expressed as [3]

$$S_{12}(f)^{(1)} = \text{Coupling} + \text{Background reflections} \qquad (4.20)$$

ii. Next, a large metal ground plane is placed at the location where the tag is intended to be placed and a second measurement, $S_{12}(f)^{(2)}$, is taken, which can be expressed as [3]

$$S_{12}(f)^{(2)} = \text{Coupling} + G_1 G_2 \left(\frac{\lambda}{4\pi \times 2r} \right)^2 \tag{4.21}$$

where r is the distance between the antenna and intended location of the tag, λ is the wavelength, and G_1 and G_2 are the transmit and receive horn antenna gains, respectively. This measurement provides information to remove the gains G_1 and G_2 introduced by the transmit and receive horn antennas.

iii. Finally, the tag is placed a at the intended location (replacing the metal plate) and a third measurement, $S_{12}(f)^{(3)}$, is carried out, which includes the response due to the tag together with the antenna gains, antenna coupling, and the background reflections. An expression for $S_{12}(f)^{(3)}$ can be written as [3]

$$S_{12}(f)^{(3)} = \text{Coupling} + G_1 G_2 \left(\frac{\lambda}{4\pi \times 2r} \right)^2 T(f) + \text{Background reflections} \tag{4.22}$$

where $T(f)$ is the transfer function that expresses the response of the tag in the frequency domain.

By the subtraction of Equation 4.20 from Equation 4.22, the unwanted antenna coupling and the background reflections can be removed successfully. This gives [3]

$$S_{12}(f)^{(3)} - S_{12}(f)^{(1)} = G_1 G_2 \left(\frac{\lambda}{4\pi \times 2r} \right)^2 T(f) \tag{4.23}$$

However, the effect of the antenna gains and path loss still remains. The second measurement is used for removing the effect of the antenna gains and path loss. This is done by first converting the $S_{12}(f)^{(2)}$ frequency-domain measurement into time domain using the inverse Fourier transform. Let $S_{12}(t)^{(2)}$ denote the time-domain signal obtained through the transform. During the experiment, since the metal ground plane was placed at a distance r away from the antennas, the time-domain signal, $S_{12}(t)^{(2)}$, should consist of predominantly two components. The first component, $f_1(t)$, is the strongest and corresponds to the coupled signal between the two antennas whereas the second component, $f_2(t)$, is much weaker and is due to the reflection of the signal at the ground plane. This second component will appear after a propagation delay of approximately $2r/c$ where c is the speed of light. Due to this propagation delay, these two components can be easily windowed out and isolated. Next, $f_2(t)$ is windowed and isolated and its Fourier transform is obtained, which provides the effect of the gains and path loss as follows [3]:

$$F\{f_2(t)\} = G_1 G_2 \left(\frac{\lambda}{4\pi \times 2r} \right)^2 \tag{4.24}$$

Therefore, using Equations 4.23 and 4.24, the tag response can be obtained as follows:

$$T(f) = \frac{S_{12}(f)^{(3)} - S_{12}(f)^{(1)}}{F\{f_2(t)\}} \tag{4.25}$$

Once the tag response, $T(f)$, is estimated, it is transformed to the time domain using the inverse Fourier transform. From this time-domain response, the source-free response of the tag is windowed and isolated. The matrix pencil algorithm is then applied on this windowed source-free tag response to obtain the pole signature of the tag. Figure 4.8 illustrates the total process involved in the complex frequency domain representation of the tag response.

4.4 CONCLUSION

The SEM expresses the late-time response of a scatter subjected to an electromagnetic pulse as a sum of damped sinusoids. The matrix pencil algorithm is a very useful mathematical tool that can be used to extract the poles and residues from a noisy sampled signal. The SEM is particularly useful in analyzing the backscatter from chipless RFID tags. This is because the completely passive chipless tags give rise to passive backscatter that are inherently of decaying nature and hence can be expressed using damped sinusoids. The chapter provided background on the SEM and a detailed explanation on the use of the matrix pencil algorithm in order to extract the complex frequency domain pole signature from a sampled noisy time-domain signal. Recent research on the use of the

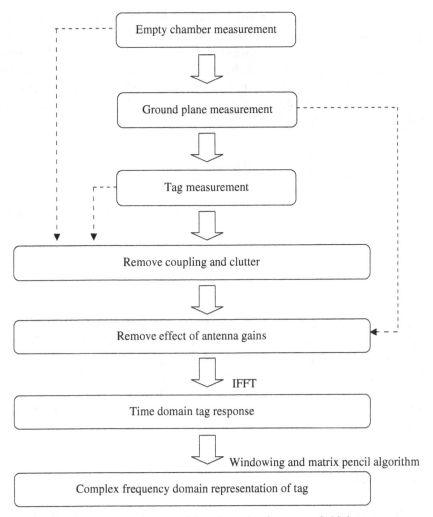

Figure 4.8   Process of obtaining the poles signature of chipless tag.

SEM for encoding information into chipless RFID tags and decoding information from them was presented. Here, the resonant properties of the tag are expressed using a unique pole signature. In addition to providing the resonance frequencies of the tag, the pole signature also provides information on how fast the backscatter from the tag is decaying. Therefore, this representation of tag response as a pole signature in the complex frequency domain provides additional information thereby giving rise to extra dimensions for data encoding, error correction, and sensing applications.

REFERENCES

1. C. Baum, "*On the singularity method for the solution of electromagnetic interaction problems,*" Air Force Weapons Lab, December 11, 1971.
2. A. Blischak and M. Manteghi, "*Pole residue techniques for chipless RFID detection,*" in *IEEE Antennas and Propagation Society International Symposium, 2009. APSURSI '9,* Charleston, SC, June 1–5, 2009, pp. 1–4.
3. A. T. Blischak and M. Manteghi, "Embedded singularity chipless RFID tags," *IEEE Transactions on Antennas and Propagation,* vol. 59, pp. 3961–3968, 2011.
4. C. Baum, "*The Singularity Expansion Method Transient Electromagnetic Fields.*" vol. 10, L. Felsen, Ed., Springer: Berlin/Heidelberg, 1976, pp. 129–179.
5. C. Baum, "The singularity expansion method: Background and developments," *IEEE Antennas and Propagation Society Newsletter,* vol. 28, pp. 14–23, 1986.
6. C. E. Baum, E. J. Rothwell, K. M. Chen, and D. P. Nyquist, "The singularity expansion method and its application to target identification," *Proceedings of the IEEE,* vol. 79, pp. 1481–1492, 1991.
7. R. S. Adve, T. K. Sarkar, O. M. C. Pereira-Filho, and S. M. Rao, "Extrapolation of time-domain responses from three-dimensional conducting objects utilizing the matrix pencil technique," *IEEE Transactions on Antennas and Propagation,* vol. 45, pp. 147–156, 1997.
8. T. K. Sarkar and O. Pereira, "Using the matrix pencil method to estimate the parameters of a sum of complex exponentials, *IEEE Antennas and Propagation Magazine,* vol. 37, pp. 48–55, 1995.
9. F. B. Hildebrand, *Introduction to Numerical Analysis,* New York: McGraw-Hill, 1956.
10. M. Van Blaricum and R. Mittra, "A technique for extracting the poles and residues of a system directly from its transient response, *IEEE Transactions on Antennas and Propagation,* vol. 23, pp. 777–781, 1975.
11. Y. Hua and T. K. Sarkar, "Matrix pencil method for estimating parameters of exponentially damped/undamped sinusoids in noise, *IEEE Transactions on Acoustics, Speech and Signal Processing,* vol. 38, pp. 814–824, 1990.
12. M. L. Van Blaricum and R. Mittra, "Problems and solutions associated with Prony's method for processing transient data, *IEEE Transactions on Electromagnetic Compatibility,* vol. EMC-20, pp. 174–182, 1978.
13. D. W. Tufts and R. Kumaresan, "Estimation of frequencies of multiple sinusoids: Making linear prediction perform like maximum likelihood," *Proceedings of the IEEE,* vol. 70, pp. 975–989, 1982.
14. R. Kumaresan and D. Tufts, "Estimating the parameters of exponentially damped sinusoids and pole-zero modeling in noise, *IEEE Transactions on Acoustics, Speech and Signal Processing,* vol. 30, pp. 833–840, 1982.

15. Y. Hua and T. K. Sarkar, "Perturbation analysis of TK method for harmonic retrieval problems," *IEEE Transactions on Acoustics, Speech and Signal Processing*, vol. 36, pp. 228–240, 1988.

16. Y. Hua and T. K. Sarkar, "On SVD for estimating generalized eigenvalues of singular matrix pencil in noise, *IEEE Transactions on Signal Processing*, vol. 39, pp. 892–900, 1991.

17. Y. Hua and T. K. Sarkar, "A perturbation property of the TLS-LP method," *IEEE Transactions on Acoustics, Speech and Signal Processing*, vol. 38, pp. 2004–2005, 1990.

18. Y. Hua and T. K. Sarkar, "On the total least squares linear prediction method for frequency estimation, *IEEE Transactions on Acoustics, Speech and Signal Processing*, vol. 38, pp. 2186–2189, 1990.

19. M. Manteghi and Y. Rahmat-Samii, "*Frequency notched UWB elliptical dipole tag with multi-bit data scattering properties*," in *2007 IEEE Antennas and Propagation Society International Symposium*, Honolulu, HI, June 9–15, 2007, pp. 789–792.

CHAPTER 5

DENOISING AND FILTERING TECHNIQUES FOR CHIPLESS RFID

5.1 INTRODUCTION

The signals backscattered from a chipless RFID tag are very weak when they reach the front end of an RFID reader. They are affected by both path loss and fading introduced by the wireless channel between the tag and the reader. The weak signals received at the reader are then affected by additive thermal noise, which further deteriorates the signal quality. The signals received at the reader in a chipless RFID system are not digitally modulated and also do not have provisions for forward error correction mechanisms like in a chipped RFID system. Therefore, the RFID reader completely relies on an analog and passive variation of a parameter in the received signal (amplitude, phase, time of arrival, etc.) to decode information bits encoded by the tag. This reliance on passive and analog transformations for representing and decoding information renders the detection process extremely vulnerable to random fluctuations of the weak received signal due to additive thermal noise and fading. Furthermore, stray echoes or reflections due to unwanted clutter present in the measurement environment also introduce ambiguity in the detection process. In addition to the errors caused by thermal noise and other

Chipless Radio Frequency Identification Reader Signal Processing, First Edition.
Nemai Chandra Karmakar, Prasanna Kalansuriya, Rubayet E. Azim and Randika Koswatta.
© 2016 John Wiley & Sons, Inc. Published 2016 by John Wiley & Sons, Inc.

signal detriments of the wireless channel, the fabrication tolerances in manufacturing chipless RFID tags also introduces errors. These errors manifest as a form of noise that shifts the resonance frequencies from their intended locations due to the random fabrication errors in the dimensions of the resonating structures in the chipless RFID tags.

In order to improve the accuracy of extracting information embedded in the received signal, the unwanted random amplitude variations due to thermal noise and clutter need to suppressed using signal processing methods. In this chapter, several signal processing methods are discussed that are useful in denoising the received backscattered signal. Noise reduction and filtering techniques especially catered for chipless RFID systems have not received a lot of attention. Only a handful of work is available in the literature that is focused on denoising signals in chipless RFID systems [1–5].

In Ref. [1], signals from a chipless RFID tag are sampled in the frequency domain and denoised using a moving average filter, which is implemented in a microcontroller-based RFID reader system. Although it is a very simple method, the filtering improves the quality of the received signal and successfully suppresses the noise contaminating it. However, when the number of points used in the moving window is increased, the sharp features of the original signal are lost due to the low pass nature of the moving average filter. Therefore, the choice of the number of points is a compromise between preserving the shape of the original signal and the amount of noise that needs to be suppressed. The authors of Ref. [2] propose a technique to reduce the noise affecting the phase of the received backscattered signal. Here, the measured phase response of a chipless RFID tag is represented as a truncated series of prolate spheroidal wave functions (PSWFs). In Ref. [5], the frequency signatures of a chipless RFID are represented using a set of orthonormal basis functions where the projection of the signature on to the signal space spanned by the basis functions filters noise present in the signatures. In Ref. [3], the backscattered signals from a chipless RFID tag are modeled using the singularity expansion method (refer Chapter 4) where the matrix pencil algorithm is used to extract information from noise-contaminated backscattered signals. In Ref. [4], the continuous wavelet transform (CWT) is used to denoise and enhance the received signal quality in a time-domain reflectometry-based chipless RFID system.

In this chapter, the aforementioned noise suppression methods will be discussed in detail. In Section 5.2, the noise filtering capabilities of the matrix pencil algorithm will be explored. Section 5.3 will present a discussion on the use of the signal space representation for noise filtering. The application of the selective spectral interrogation (SSI) method for frequency-domain chipless RFID tag for enhancing signal peaks and suppressing affects of unwanted noise and clutter is discussed in Section 5.4. The use of wavelet-based signal processing for denoising and improving signal quality is presented in Section 5.5. The chapter conclusions are drawn in Section 5.6.

5.2   MATRIX PENCIL ALGORITHM-BASED FILTERING

The backscatter produced by a chipless RFID tag when it is illuminated by a radio frequency pulse has a transitory and decaying nature. In Chapter 4, it was shown that these decaying signals can be best approximated using the singularity expansion method where the backscattered signal is expressed as a sum of decaying sinusoids (refer Eq. 5.1). A mathematical tool, the matrix pencil algorithm, was also introduced that can extract information from these signals. By using the matrix pencil algorithm, it is possible to extract both the resonant frequencies and the rates of decay or the damping coefficients associated with the received tag backscattered signals, whereas through the use of a fast Fourier transform (FFT), only the resonant frequencies can be extracted. The total least-squares matrix pencil (TLSMP) method [6] is a variant of the matrix pencil algorithm, which is capable of combating the detrimental effects of noise contamination in a signal.

Let y be the noise-contaminated backscattered signals received from a chipless tag:

$$y(t) = \sum_{i=1}^{P} a_i e^{-\sigma_i t} \cos(\omega_i t + \phi_i) + n(t) = \sum_{i=1}^{M} R_i e^{s_i t} + n(t) \qquad (5.1)$$

where a_i and ϕ_i are the amplitude and phase, σ_i is the damping factor, ω_i is the angular frequency of the ith damped sinusoid, P is the number of sinusoids, $s_i = -\sigma_i + j\omega_i$ are the poles of the signal, R_i are the complex amplitudes or residues associated with each pole, $M = 2P$ is the number of poles, and $n(t)$ is the additive noise contaminating the signal. The additive noise is generally modeled as additive white Gaussian noise in research. Let y_k be the kth sample of the time-domain received signal $y(t = k\Delta t)$ where Δt is the sampling interval. Following are the steps involved in applying the TLSMP algorithm on a noisy set of samples y_k to extract the poles and residues of the signal.

STEP 1: Choose N samples and a matrix $[Y]$ is formed as follows:

$$[Y] = \begin{bmatrix} y_0 & y_1 & \cdots & y_L \\ y_1 & y_2 & \cdots & y_{L+1} \\ \vdots & \vdots & \ddots & \vdots \\ y_{N-L-1} & y_{N-L} & \cdots & y_{N-1} \end{bmatrix}_{(N-L) \times (L+1)} \tag{5.2}$$

where L is be chosen to be between $N/3$ and $N/2$ [7–9].

STEP 2: Perform singular value decomposition or SVD of $[Y]$ and obtain the unitary matrices $[U]$ and $[V]$ and the diagonal singular value matrix $[\Sigma]$:

$$[Y] = [U][\Sigma][V]^H \tag{5.3}$$

STEP 3: Estimate new matrix $[V']$ from $[V]$; first, select M' dominant singular values from the original singular value matrix $[\Sigma]$ and discard all other spurious singular values, which are due to noise. Produce the new matrix $[V']$ by removing the eigenvectors from $[V]$, which correspond to the removed spurious singular values of $[\Sigma]$ (Eq. 4.17).

STEP 4: Produce matrices $[V_0']$ and $[V_0']$ using $[V']$; $[V_0']$ is the matrix obtained by removing the last column from $[V']$, and $[V_1']$ is the matrix obtained by removing the first column from $[V']$.

STEP 5: Solve for the ordinary eigenvalues of the matrix $B = \{[V_0']^H\}^\dagger [V_1']^H$. The nonzero eigenvalues of $\{[V_0']^H\}^\dagger [V_1']^H$ will be equal to the z_i parameters of the signal, where $z_i = e^{s_i \Delta t} = e^{(-\sigma_i + j\omega_i)\Delta t}$ and $[A]^\dagger$ is the Moore–Penrose pseudoinverse of $[A]$. The complex pole singularities can be obtained from z_i parameters as follows:

$$s_i = -\sigma_i + j\omega_i = \frac{\ln z_i}{\Delta t} \tag{5.4}$$

STEP 6: Solve the following equation to obtain the residues

$$\begin{bmatrix} y_0 \\ y_1 \\ \vdots \\ y_{N-1} \end{bmatrix} = \begin{bmatrix} 1 & 1 & \cdots & 1 \\ z_1 & z_2 & \cdots & z_M \\ \vdots & \vdots & \ddots & \vdots \\ z_1^{N-1} & z_2^{N-1} & \cdots & z_M^{N-1} \end{bmatrix} \begin{bmatrix} R_1 \\ R_2 \\ \vdots \\ R_M \end{bmatrix} \tag{5.5}$$

Figure 5.1 shows a simulation result to demonstrate the noise filtering capability of the TLSMP algorithm. The signal shown in Figure 5.1a is

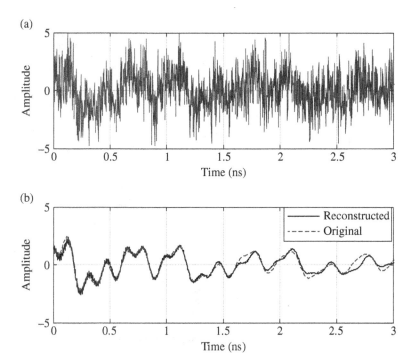

Figure 5.1 (a) Noise-contaminated signal (SNR = −6 dB). (b) Original signal and reconstructed signal.

heavily affected by noise to the extent that it is not possible to make out the shape of the original signal. After application of the TLSMP algorithm on the noisy samples, the poles, $s_i = -\sigma_i + j\omega_i$, and the residues, R_i, of the original signal are estimated. Using these estimated poles and residues, an estimate $\tilde{x}(t)$ of the original signal can be reconstructed as follows:

$$\tilde{x}(t) = \sum_{i=1}^{\tilde{M}} \tilde{R}_i e^{\tilde{s}_i t} \tag{5.6}$$

where $\tilde{s}_i$ and $\tilde{R}_i$ are the estimates of the poles and residues and $\tilde{M}$ is the estimate of the number of poles. From Figure 5.1b, it is clear that this reconstructed estimate closely follows the contour of the original signal. In the simulation, the noisy signal was sampled with a Δt of 2 ps and for the TLSMP algorithm $N = 1000$ and $L = 500$. It should be noted that the TLSMP algorithm is more computationally intensive than an FFT operation. Figure 5.2 shows the complex conjugate pole pairs of

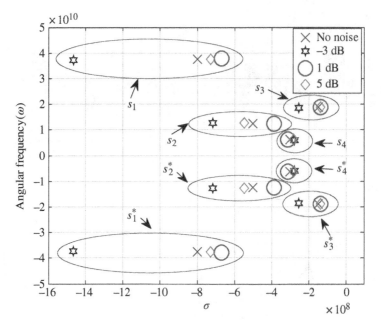

Figure 5.2 Poles of the signal for different SNR.

the signal for different signal-to-noise ratios (SNR). From the figure, it is clear that the estimated angular frequency does not show a lot of error, whereas the damping coefficient shows errors when the SNR drops. The errors are more pronounced as the damping factor gets smaller.

The research detailed in Ref. [3] shows that the autocorrelation function of a signal can be used to extract the poles of the signal. The autocorrelation function of a signal $y(t)$ is defined as follows:

$$R_{yy}(\tau) = \int_{-\infty}^{\infty} y(t) y^*(t-\tau) dt, \tag{5.7}$$

where a^* is the complex conjugate of a. The Laplace transform of $y(t)$ reveals its poles s_i and residues R_i as follows:

$$Y(s) = L\{y(t)\} = \sum_{i=1}^{M} \frac{R_i}{s - s_i} \tag{5.8}$$

The Laplace transform of $R_{yy}(\tau)$ can be computed as [3]

$$
\begin{aligned}
L\{R_{yy}(\tau)\} &= Y(s)Y^*(s) \\
&= \left(\sum_{i=1}^{M}\frac{R_i}{s-s_i}\right)\left(\sum_{i=1}^{M}\frac{R_j^*}{s-s_j^*}\right) \\
&= \sum_{i=1}^{M}\sum_{j=1}^{M}\frac{R_i R_j^*}{(s-s_i)(s-s_j)} \\
&= \sum_{i=1}^{M}\left(\sum_{j=1}^{M}\frac{R_i R_j^*}{(s-s_j^*)}\right)\frac{1}{s-s_i} - \sum_{i=1}^{M}\left(\sum_{j=1}^{M}\frac{R_i R_j^*}{(s-s_j^*)}\right)\frac{1}{s-s_j^*} \\
&= L\{R_{yy}^+(\tau)\} + L\{R_{yy}^-(\tau)\}
\end{aligned}
\tag{5.9}
$$

where $R_{yy}^+(\tau)$ and $R_{yy}^-(\tau)$ are the parts of the autocorrelation function in the positive and negative sides of the τ axis. It is clear from (5.9) that $R_{yy}^+(\tau)$ has the exact same set of poles as $y(t)$. The convolution operation performed in (5.7) to calculate $R_{yy}(\tau)$ removes a considerable amount of noise from the signal $y(t)$. Therefore, the poles of the signal can be calculated with more accuracy amidst noise by applying the TLSMP algorithm on the computed $R_{yy}^+(\tau)$ than by directly applying it on the sampled data. Using the calculated poles, the residues can be computed by solving (5.5).

5.3 NOISE SUPPRESSION THROUGH SIGNAL SPACE REPRESENTATION

The signal space representation of chipless RFID tag signatures [5, 10], discussed in Chapter 2, requires a set of orthonormal basis functions calculated using all the possible frequency signatures of the chipless RFID system. Through the use of these orthonormal basis functions, a mathematical L-dimensional signal space is constructed to represent the different tag frequency signatures. In the detection process, a noise-contaminated frequency signature is projected on to the signal space and is compared against existing constellation points in the space to extract information contained in the noisy signature. This projection is performed by taking the inner product or dot product with each of the orthonormal basis functions that serve as the unit vectors along the axes of the L-dimensional signal space. The projection of the noisy detected signature on to the signal space filters out most of the noise contamination. The steps involved in the signal space representation of tag frequency signatures are summarized below.

STEP 1: Form a matrix $[H]$ using all possible frequency signatures of the b-bit chipless RFID system and perform SVD of $[H]$:

$$[H] = [\Phi][\Sigma][\Theta]^{H} \tag{5.10}$$

where $[\Phi]$ and $[\Theta]$ are unitary matrices and $[\Sigma]$ is a diagonal matrix containing singular values.

STEP 2: Choose the first L orthonormal vectors φ_i from the matrix $[\Phi]$ that correspond to the first L significant singular values of $[\Sigma]$. These L vectors will serve as the basis functions for the signal space representation.

STEP 3: Obtain inner product values between the noisy tag signature, $\mathbf{y}$, and the basis functions and calculate the noisy signal point $\mathbf{r} = [\mathbf{r}_1,...,\mathbf{r}_L]$ in the signal space corresponding to the noisy tag signature:

$$\mathbf{r} \equiv \left[\langle \mathbf{y}, \varphi_1 \rangle, \langle \mathbf{y}, \varphi_2 \rangle, ..., \langle \mathbf{y}, \varphi_L \rangle \right] \tag{5.11}$$

STEP 4: Use minimum distance decoding [11] to extract information contained in the noisy received signal point.

Figure 5.3 presents an overview of the signal space representation of a 3-bit chipless RFID system. The 3-bit chipless RFID system gives rise to eight possible tag signatures. The SVD process produces four orthonormal basis functions that can be used to represent any of the eight possible tag signatures. Three of these basis functions are used in forming a three-dimensional signal space (three basis functions are selected for the ease of visualizing a three-dimensional space as opposed to a four-dimensional space) in which the eight signatures are represented as eight points.

The noise mitigation achieved through this approach is illustrated using a simulation result in Figure 5.4. For this illustration, the frequency signature $\mathbf{h}_3$ shown in Figure 5.3 is contaminated with additive Gaussian noise so that the SNR is 1 dB (shown in Fig. 5.4a). In this example, the noise contamination is in the frequency domain as opposed to the time domain as in Figure 5.1. The inner products between this noisy signature and the four basis functions $\mathbf{u}_1$, $\mathbf{u}_2$, $\mathbf{u}_3$, and $\mathbf{u}_4$ (shown in Fig. 5.3) are computed. These inner products are then used together with the basis functions in order to reconstruct the original frequency signature as follows:

$$\mathbf{r} = \sum_{i=1}^{4} \langle \mathbf{y}, \varphi_i \rangle \varphi_i \tag{5.12}$$

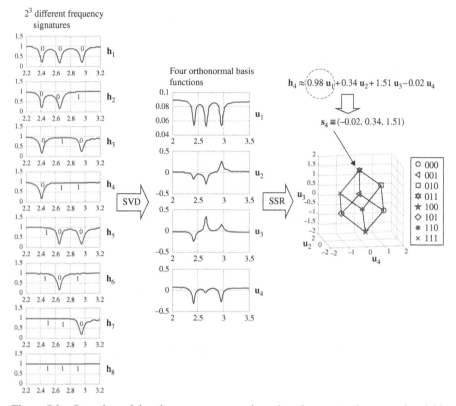

Figure 5.3 Overview of signal space representation of tag frequency signatures in a 3-bit chipless RFID system.

where $y = h_3 + n$ is the noise-contaminated frequency signature, n is the noise, and r is the reconstructed signal. The original frequency signature, h_3, and the reconstructed frequency signature, r, computed using (5.12) are shown in Figure 5.4b. It is clear that the noise has been filtered out in the reconstructed signature. The signal space representation of this noisy signature is shown in Figure 5.5. The signal point R corresponding to the noisy signature is closer to the point corresponding to the tag signature h_3. Therefore, the information contained in the signature can be correctly decoded as "010."

Here, the amplitude spectrum of the frequency signature of a chipless RFID tag was represented using a set of orthonormal basis functions. A similar study is presented in Ref. [2] where the phase spectrum of a chipless RFID tag is represented using a truncated series of PSWFs. PSWFs are an orthonormal basis for a vector space formed by band-limited functions having a given bandwidth. Because the

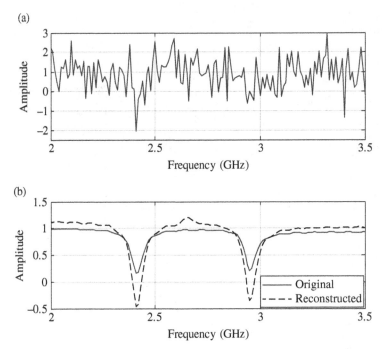

Figure 5.4   (a) Noisy tag frequency signature (SNR = 1 dB). (b) Original frequency signature and reconstructed frequency signature.

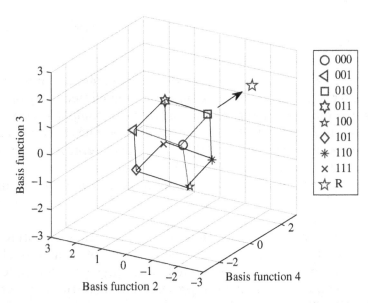

Figure 5.5   Reconstructed frequency signature plotted in the three-dimensional signal space as a point *R*.

backscatter produced by chipless tags is also band limited, the PSWFs are a good set of orthonormal basis functions for the purpose of representing the backscatter. The principle behind the noise filtering is the same as that presented earlier for the representation of the backscattered amplitude spectrums in a signal space. In this case, since noise contaminating the backscattered signal is not band limited and spread across a wide bandwidth when it is projected into the vector space defined by the set of band-limited orthonormal basis functions, most of the noise is filtered out.

5.4 SSI

In the UWB impulse interrogation of frequency signature-based chipless RFID tags, the backscatter received is sampled in the time domain where the unwanted coupled signal or rejection signal and the structural mode backscatter is identified. By using a windowing function, only the backscatter that contains the resonance information or frequency signature of the chipless tag is filtered out from the time-domain signal.

For the purpose of interrogating the chipless tags, a broadband pulse is required. A common technique to produce such a broadband pulse is through the use of a modulated Gaussian pulse $x(t)$, which can be expressed as [12]

$$x(t) = A_0 \cos(2\pi f_c t) \exp\left(-\frac{(t-\mu)^2}{2\sigma^2}\right) \tag{5.13}$$

where A_0 is the amplitude and f_c is the frequency of the sinusoidal carrier frequency. The parameters μ (mean) and σ (standard deviation) define the shape of the time-domain Gaussian pulse. The bandwidth is determined by the width of the pulse in the time domain, which is controlled by the σ parameter. Figure 5.6 shows a modulated Gaussian pulse having a 20 dB bandwidth of 6 GHz spanning from 2 to 8 GHz with a center frequency of 5 GHz where $A_0 = 1$, $f_c = 5$ GHz, $\mu = 0.6$ ns, and $\sigma = 0.114$ ns. It is clear from the Figure 5.6b that the interrogation pulse has a Gaussian-shaped amplitude spectrum. Figure 5.7 shows the measured frequency signatures of tags placed at different distances from a reader antenna in an anechoic chamber environment [12]. These frequency signatures were calculated by taking the Fourier transform of

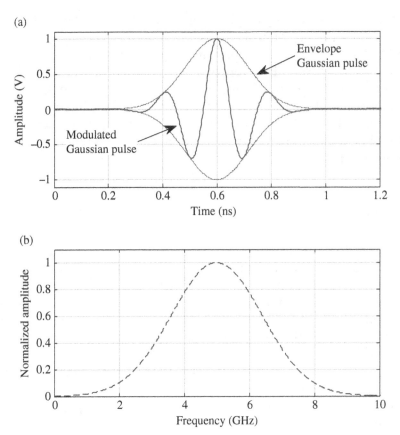

Figure 5.6 (a) Shape of the modulated Gaussian interrogation pulse in the time domain. (b) Amplitude spectrum of the interrogation pulse.

the windowed antenna mode backscatter as explained in Chapter 3. The chipless RFID system considered in the experiment is a 4-bit system were each bit is represented using the presence of a unique resonance frequency. The resonant frequencies $f_1 = 4.6\,\text{GHz}, f_2 = 5.1\,\text{GHz},$ $f_3 = 5.5\,\text{GHz}$, and $f_4 = 6.2\,\text{GHz}$ were used. A chipless tag containing two $(f_1$ and $f_3)$ of these four resonance frequencies was used for obtaining the measurement results shown in Figure 5.7.

From the results shown in Figure 5.7, it is clear that as the distance increases, the detection of the resonant peaks becomes more difficult and ambiguous due to the presence of spurious peaks, particularly at the higher frequencies (Fig. 5.7c). These spurious peaks in the frequency signatures might result due to noise and clutter present in the environment. Since the interrogation pulse has a Gaussian-shaped amplitude spectrum, the higher frequencies have lower amplitude than the center

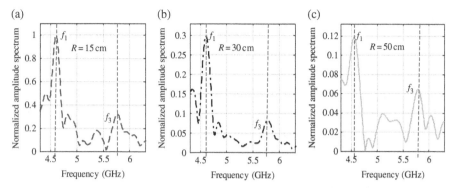

Figure 5.7 Measured frequency spectra of chipless tag having two resonances (4.6 and 5.8 GHz) placed $R = 15$, 30, and 50 cm in front of the reader antenna [12].

frequencies. This further aggravates the detection problem since the resonant peaks corresponding to the higher frequencies are smaller compared to the center frequencies rendering them more susceptible to noise. Because of the uneven heights of the resonant peaks, it is not possible to use a fixed threshold level for deciding whether a peak in the spectral signature is due to a resonant frequency of the tag. Therefore, additional signal processing is required for the accurate detection of the information contained in the spectral signature of the chipless tag. In Ref. [12], a novel technique termed SSI is proposed for the detection of information bits in a context like this. Here, the response from the tag for a set of interrogation pulses, $s_i(t)$, $i = 1,\ldots,4$, is used to determine which resonant frequencies are present in the tag. These interrogation pulses are defined as follows:

$$s_i(t) = A_i \cos(2\pi f_i t) \exp\left(-\frac{(t-\mu_2)^2}{2\sigma_2^2}\right), \quad i = 1,\ldots,4 \qquad (5.14)$$

where $A_i = 1$, $i = 1,\ldots,4$. Here, the carrier frequency of the modulated Gaussian pulse is chosen to be the resonant frequencies of the tag, f_i, $i = 1,\ldots,4$. The parameter σ_2 of the Gaussian pulse is larger than that of (5.13), which makes this pulse more wider in the time domain and less broadband with more energy concentrated around f_i. Figure 5.8 shows the amplitude spectra of the four interrogation pulses expressed in (5.14). Each pulse has a 20 dB bandwidth of approximately 2.2 GHz, where $A_i = 1$, $\mu_2 = 0.6$ ns, and $\sigma_2 = 0.3$ ns.

The principle behind the operation of this new method is as follows. When the tag is illuminated with the pulse $s_j(t)$, $i = j$, if the tag contains

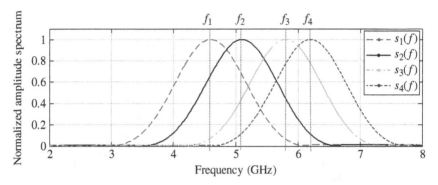

Figure 5.8 Amplitude spectra of the interrogation pulses $s_i(t)$, $i = 1,...,4$.

a resonant frequency at f_j, the amplitude spectrum of the antenna mode backscatter should show a peak at f_j. This peak will be significantly larger than all the other peaks corresponding to the other resonant frequencies of the tag or spurious peaks due to noise and clutter. This is because the interrogation pulse, $s_j(t)$, $i = j$, contains more energy at the vicinity of f_j than all the other resonance frequencies. As a result, the tag resonance at f_j becomes more prominent if it exists. If such a maxima in the amplitude spectrum is not seen, it implies that the tag does not contain a resonance at f_j. For an example, consider the frequency signature shown in Figure 5.7c, which is contaminated with some spurious spectral peaks causing an ambiguity in determining the actual resonances present in the chipless tag. This frequency signature was obtained by interrogating the tag using the pulse defined in (5.13), which has a wider bandwidth. Let us now consider this tag at the same position with the new SSI method to extract the information contained in it. Figure 5.9 shows the four different frequency spectra of the antenna mode backscatter of the tag that correspond to a different interrogation pulse $s_i(t)$, $i = 1,...,4$. From the figure, it is clear that when $s_1(t)$ and $s_3(t)$ are used to interrogate the tag, the maximum amplitude occurs at the corresponding resonant frequencies of f_1 and f_3, whereas when the pulses $s_2(t)$ and $s_4(t)$ are used, the maximum amplitude does not occur at the respective resonances of f_2 and f_4. This confirms that the tag does not contain any resonances at frequencies f_2 and f_4 and that it only contains the resonances at frequencies f_1 and f_3. Clearly, the new approach has removed the ambiguity involved in the detection due to spurious peaks where the detection is now performed using a peak detection method without the need for a fixed threshold level.

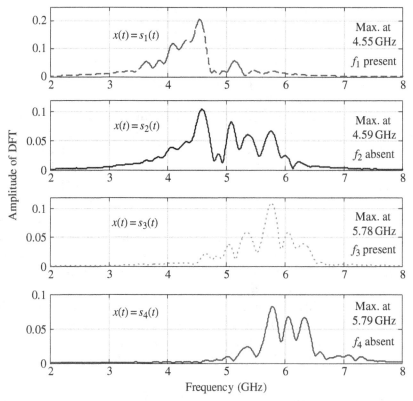

Figure 5.9 Amplitude spectrums of the windowed antenna mode backscatter from the tag (having resonances at f_1 and f_3) when different interrogation pulses are used [12].

5.5   WAVELET-BASED FILTERING OF NOISE

The CWT of a signal $s(t)$ at a scale a and a translation τ is defined as [4]

$$S(a,\tau) = \int_{-\infty}^{+\infty} s(t)\frac{1}{\sqrt{a}}\psi *\left(\frac{\tau-t}{a}\right)dt \qquad (5.15)$$

where $\psi(t)$ is a function that is continuous in both time domain and frequency domain and is called the mother wavelet. The CWT compares the shape of the signal $s(t)$ with the shape of the wavelet for a given translation in time and a scaling factor. The translation provides the ability to localize the wavelet shape in the time domain. The scaling factor dilates or contracts the wavelet shape that is being compared with the signal $s(t)$. Scaling of the wavelet affects its frequency content.

By performing CWT for a range of translations and scaling factors, a time–frequency representation of a signal can be obtained.

The research reported in Ref. [4] uses the CWT to enhance the SNR of the backscattered signal in a time-domain reflectometry-based chipless RFID system. Here, the signal being transformed or being compared against the wavelet is the noisy backscattered signal from a chipless tag. The transmitted interrogation pulse has a Gaussian shape and it undergoes a derivative effect when it is being transmitted and received via antennas. Therefore, the received backscatter will have a shape of a higher-order derivative of a Gaussian pulse. Therefore, a Gaussian-shaped wavelet would be suitable for the CWT. In a time-domain reflectometry-based chipless RFID tag, the information is stored using the time of arrival information of the backscattered pulse. Using the CWT, it is possible to localize and identify the presence of the backscattered Gaussian pulse amidst noise in the total backscattered signal. Essentially, when different scaling and translations are being used, the CWT works as a matched filter optimized and matched to the received pulse, which is contaminated by noise and shifted in time due to a propagation delay between the tag and reader. The experimental results presented in Ref. [4] report that a read range can be enhanced up to 2 m through the CWT-based signal enhancement.

5.6 CONCLUSION

Backscattered signals from chipless RFID tags are very weak, and the detection process at the RFID reader is extremely susceptible to noise. This is because the information is represented as analog variations as opposed to a modulated digital steam of data as in conventional wireless communication of information where error correction schemes can be used to detect the presence of errors and discard erroneous data packets.

The TLSMP algorithm and the CWT are two approaches to filtering noise present in the signal in the time domain. It is possible to reconstruct a noise-free version of a noisy signal by using poles and residues estimated through the TLSMP algorithm. However, the algorithm is computationally intensive than a conventional Fourier transform-based analysis. The CWT enables the creation of a time–frequency representation of the backscattered signal, which is constructed by comparing the received backscattered signal with a known wavelet function for different scaling and translation factors. This representation enables the accurate localization of the backscattered pulses in the time domain using a noisy signal.

The signal space representation uses an orthonormal set of basis functions to represent chipless RFID frequency signatures in a signal space. The projection of the noisy received frequency signature on to the signal space filters out a considerable amount of noise affecting the signal. SSI is a technique used to enhance the effect of actual resonances in the chipless RFID tag in order to remove the ambiguity caused due to spurious spectral peaks due to noise and clutter.

Through these noise mitigation techniques, it is possible to enhance the quality of the received signal at the RFID reader so that more accurate decisions can be made in the process of extracting information bits encoded in the backscattered signals.

REFERENCES

1. R. Koswatta and N. C. Karmakar, "*Moving average filtering technique for signal processing in digital section of UWB chipless RFID reader*," in *2010 Asia-Pacific Microwave Conference Proceedings (APMC)*, Yokohama, December 7–10, 2010, pp. 1304–1307.

2. W. Dullaert, L. Reichardt, and H. Rogier, "Improved detection scheme for chipless RFIDs using prolate spheroidal wave function-based noise filtering," *IEEE Antennas and Wireless Propagation Letters*, vol. 10, pp. 472–475, 2011.

3. M. Manteghi, "*A novel approach to improve noise reduction in the Matrix Pencil Algorithm for chipless RFID tag detection*," in *IEEE Antennas and Propagation Society International Symposium (APSURSI), 2010*, Toronto, ON, July 11–17, 2010, pp. 1–4.

4. A. Lazaro, A. Ramos, D. Girbau, and R. Villarino, "Chipless UWB RFID tag detection using continuous wavelet transform," *IEEE Antennas and Wireless Propagation Letters*, vol. 10, pp. 520–523, 2011.

5. P. Kalansuriya, N. Karmakar, and E. Viterbo, "A novel approach in the detection of chipless RFID," in *Chipless and Conventional Radio Frequency Identification: Systems for Ubiquitous Tagging*, N. Karmakar, Ed., Hoboken: IGI Global, 2012, pp. 218–233.

6. R. S. Adve, T. K. Sarkar, O. M. C. Pereira-Filho, and S. M. Rao, "Extrapolation of time-domain responses from three-dimensional conducting objects utilizing the matrix pencil technique," *IEEE Transactions on Antennas and Propagation*, vol. 45, pp. 147–156, 1997.

7. Y. Hua and T. K. Sarkar, "On SVD for estimating generalized eigenvalues of singular matrix pencil in noise," *IEEE Transactions on Signal Processing*, vol. 39, pp. 892–900, 1991.

8. Y. Hua and T. K. Sarkar, "A perturbation property of the TLS-LP method," *IEEE Transactions on Acoustics, Speech and Signal Processing*, vol. 38, pp. 2004–2005, 1990.

9. Y. Hua and T. K. Sarkar, "On the total least squares linear prediction method for frequency estimation," *IEEE Transactions on Acoustics, Speech and Signal Processing*, vol. 38, pp. 2186–2189, 1990.

10. P. Kalansuriya, N. Karmakar, and E. Viterbo, "*Signal space representation of chipless RFID tag frequency signatures*," presented at the IEEE Global Communications Conference, GLOBECOM 2011, Houston, TX, December 5–9, 2011.

11. S. Haykin, *Communication Systems*, 5 ed., New York: John Wiley & Sons, Inc., 2009.

12. P. Kalansuriya, N. C. Karmakar, and E. Viterbo, "On the detection of frequency-spectra-based chipless RFID using UWB impulsed interrogation," *IEEE Transactions on Microwave Theory and Techniques*, vol. 60, pp. 4187–4197, 2012.

CHAPTER 6

COLLISION AND ERROR CORRECTION PROTOCOLS IN CHIPLESS RFID

6.1 INTRODUCTION

The low-cost alternatives of costly chipped tags, which are the chipless RFID tags, have generated significant research interest in mass deployment of low-cost item tagging. The different types of readers and their conventional and nonconventional detection methods for tag identification have been discussed in the previous chapters. However, the mass deployment of chipless RFID introduces many new challenges to the system. Item-level tagging and detection in mass deployment with chipless tags will bring a fully new set of technical specification requirements. Among the different challenges and issues, the tag collision, multiple tag reading, and data integrity are some key challenges that need to be addressed properly for improving the reliability, authenticity, and increasing acceptance of chipless RFID systems in various industry applications. There are different prospective application areas where chipless RFID is creating its way such as airport baggage handling, smart libraries, smart shelves containing RFID tagged items, and tracking of objects. In these applications, simultaneous transmission from multiple tags is very likely to occur. This causes

Chipless Radio Frequency Identification Reader Signal Processing, First Edition.
Nemai Chandra Karmakar, Prasanna Kalansuriya, Rubayet E. Azim and Randika Koswatta.
© 2016 John Wiley & Sons, Inc. Published 2016 by John Wiley & Sons, Inc.

data collision and the reader is unable to decode the tag ID correctly. To address the collision problem, RFID systems are installed with different anticollision protocols. Unfortunately, the established anticollision protocols used in conventional RFID systems cannot be applied straight away to chipless RFID. The reason is that chipless tags are void of active elements like ASIC and battery, which makes them incapable of employing signal processing algorithms for anticollision protocols within the tag. Hence, the burden of collision detection, avoidance, and multi-access methods lies within the reader part only. The data integrity and reliability of RFID system is another key issue in RFID system. In conventional RFID system, various error correction protocols and detection techniques are used for improving the data integrity. These methods are used in single tag and multi-tag environment [1] to accommodate decoding error in tag identification. This chapter gives an insight about (i) the available anticollision algorithms used for conventional and SAW chipless tags, (ii) their effectiveness in frequency-domain chipless RFID systems, (iii) the probable approach for collision detection, (iv) multiple access for chipless RFID system, and finally (v) a method for data integrity and reliability improvement of the chipless RFID system using on block coding. The detailed approaches that are being investigated for multiple access in chipless RFID system are elaborated in the following two chapters. Figure 6.1 gives an overview of the

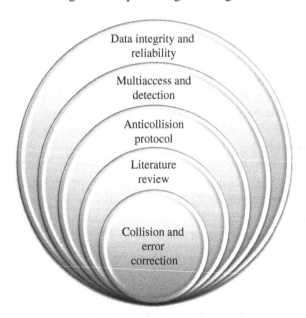

Figure 6.1 Organization of the chapter.

organization of the chapter. As can be seen in the figure, a comprehensive review of the topic has been described first. Various anticollision protocols used in the conventional chipped tags have been described. As mentioned before, only SAW tags have gained commercial success due to its compatibility with the existing infrastructure. The anticollision protocol is also revisited in this context. However, the passive chipless tags or radar arrays have no such anticollision protocols existing thus far. Therefore, we have proposed a few multi-access techniques for the chipless RFID that are usually used in telecommunication systems. They are spatial diversity (space division) multiple access (SDMA), frequency division multiple access (FDMA), time division multiple access (TDMA), and code division multiple access (CDMA). Finally, data integrity and reliability for chipless RFID tags are presented. In this space, linear block code (LBC) for frequency-domain chipless RFID tags is proposed as a viable method for collision avoidance. LBC improves reliability and data integrity in other chipless RFID systems as well.

6.2   RFID SYSTEM AND COLLISION

An RFID system usually includes a reader and multiple RFID tags. In many applications within the interrogation zone of one reader, multiple tags are present. Two main forms of communication exist in an RFID system. First one is the transmission of an interrogation signal to the interrogation zone of a reader. The transmitted signal is received by multiple tags in the interrogation zone. The tags modulate the signal and send back data to the reader. A block diagram of such a system is shown in Figure 6.2. In communication systems, the channel is subdivided to multiple users in a way that their data streams do not interfere with each other. But in passive (both chipped and chipless) RFID systems, the receiver section of the reader is available as a channel for transferring data from tags to a reader. Therefore, the responses from various tags interfere with each other and make it challenging for the reader to decode the tag's IDs correctly [2]. This scenario is termed as collision in RFID nomenclature.

Collision may have different forms in an RFID system based on the interaction between readers and tags [3]. They can be grouped as (i) reader–reader collision, (ii) reader–tag collision, and (iii) tag–tag collision. The following are the detailed description of the different collisions in RFID systems.

6.2.1   Reader–Reader Collision

Reader–reader collision occurs when a tag remains in the interrogation zone of multiple readers simultaneously. It receives the interrogation signal from multiple readers and tries to respond back to the queries. Hence, the tag may fail to send the response to any reader at all, or a wrong decoding may occur due to the collision (Fig. 6.3a).

6.2.2   Reader–Tag Collision

The reading zones of two or more readers may overlap and interfere with each other. Therefore, the reader may receive the signal from a tag and from another reader simultaneously. This scenario is termed as reader–tag collision (Fig. 6.3b).

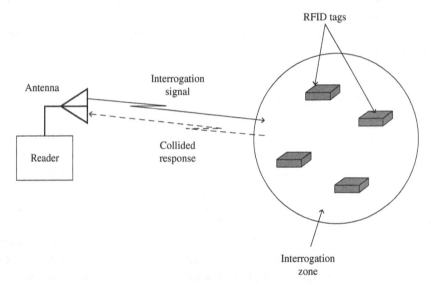

Figure 6.2   Interrogation by the reader and simultaneous reception from multiple tags.

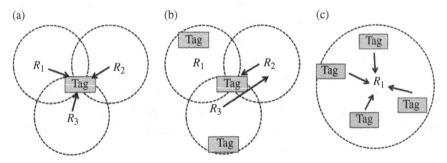

Figure 6.3   (a) Reader–reader collision, (b) reader-tag collision, and (c) tag–tag collision.

6.2.3 Tag–Tag Collision

This is the most common type of collision problem in the RFID system. Figure 6.3c explains the problem. Here, many individual tags are transmitting their ID to a reader simultaneously. When the reader tries to identify the tag ID, it fails to do so as the overlapping response from multiple tags may cancel each other or may lead to decode a wrong ID.

6.3 APPLICATIONS THAT INVOLVE MULTIPLE TAGS

Automatic identification (Auto-ID) is a key term in today's world with an aim to tag each item with a unique ID so that the tagged items can be monitored and tracked immediately without frequent human intervention. This extended capacity emerges the scope of RFID technology to a huge extent. According to IDTechEx, the potential market scope of passive RFID includes drug, healthcare, retail, postal, library, document, smart card, smart tickets, air baggage, animal tracking and vehicles [4–6]. In many applications, a single-read RFID tag may serve the demand. However, with the expansion of RFID application, there are several applications where simultaneous responses from multiple proximity tags may occur. The passive tags (chipped and chipless) are unable to sense the environment and cannot detect when collision occurs. The tags need to be read efficiently when multiple tags are in close proximity to each other. If the reader is not furnished with collision detection methods, a wrong detection may result. However, in some applications, simultaneous reading of multiple tags in the same interrogation zone of a reader is highly desirable. Hence, collision resolution methodology is a challenging and potential area of research for improving reliability of single tag reading as well as creating multi-access scope for RFID system. The successful reading of multiple tags in a complex environment is termed as throughput. The prime objective is to improve the throughput of an RFID system. This chapter is therefore about the throughput improvement implementing various anticollision protocols, error correction coding, and data integrity techniques.

Some applications of multi-tag reading are shown in Figures 6.4, 6.5, 6.6, and 6.7. Figure 6.4 shows an RFID-enabled smart shelve, which is an excellent example of multiple tag reading. Smart shelves allow inventory check wirelessly and without any human intervention. A specific example of an RFID-enabled smart shelf for a library database management system [5] is shown in Figure 6.4 when books are tagged with RFID tags. As the tags are remaining in very close proximity to each other, multiple tags may reply to the reader's query and collision

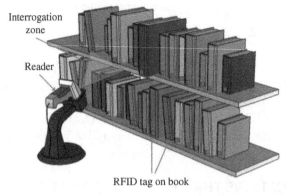

Figure 6.4 Library with smart tagging system [5].

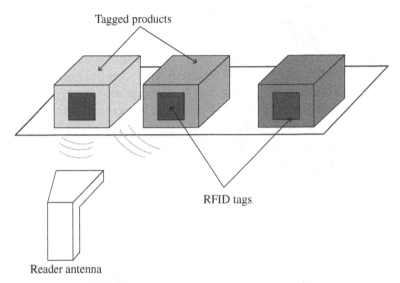

Figure 6.5 Tagged item on a conveyer belt [6].

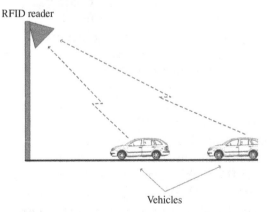

Figure 6.6 RFID for vehicle entry tracking [7].

Figure 6.7   RFID in warehouse management [4].

occurs. If the reader is provided with anticollision and multi-access capability, false detection due to collision can be avoided as well as multiple tags can be reliably read simultaneously. Simultaneous reading of several tags may also require in airport luggage tracking on conveyer belt (Fig. 6.5) [6], vehicle tracking while entering to a permit zone, toll collection (Fig. 6.6) [7], and warehouse management (Fig. 6.7) [4].

The requirement of multipleaccess in RFID systems is becoming prominent day by day. However, in the emerging chipless RFID systems, the tags are low-cost printed structures, and modification of tag for anticollision is not realistic. Therefore, the collision detection and multiple tag reading algorithms need to be implemented in the reader only. It is obvious that if the chipless RFID system can be modified for simultaneous detections of multiple tagged items, it can touch many new application areas.

To address the collision problem and multiple access with an RFID system, whether chipped or chipless, the main tasks are shown in Figure 6.8. It involves three main steps. The first one is to detect the event of collision. The reader needs to sense simultaneous responses from multiple tags that are happening within the interrogation zone. The next step is to detect the number of tags colliding during the identification period. It is necessary for the reader to have the number of collided tags that enables it to adjust its frame size or total reading time. The next step is to read the tags individually. This can be done by several methods. If the tags are capable of sensing the collision and

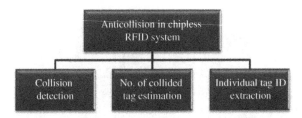

Figure 6.8   Detailed steps for multiple access in RFID system.

communicate among themselves, they can respond one by one in each time slot so that identification can be done without collision. Another way is to interrogate the tags one by one by the reader by first sending a sleep signal to all the tags and then sending a wake-up signal to turn on the tags one by one.

However, it is challenging to address the above three tasks by using any single anticollision method for collision avoidance in chipless RFID system. Therefore, in most of the literature, it is proposed to use a combination of algorithms for collision avoidance [8, 9] and consequently throughput improvement. The anticollision algorithms required for different application areas may also vary. For example, in some applications such as vehicle entry and toll collection, SDMA may serve the purpose. But in complex warehouse environment, as shown in Figure 6.7, SDMA together with CDMA may provide more reliability and robust identification. Therefore, it may be concluded that the anticollision algorithms are application specific to some extent.

6.4   ANTICOLLISION ALGORITHM IN CHIPPED RFID TAGS

Among the three types of collision described in the previous section, *tag–tag* collision is the most common one in large RFID systems where one reader and multiple tags are installed. Therefore, most of the research related to anticollision focuses on *tag–tag* collision problem in RFID. Numerous methods have been developed to solve the collision problem in the wireless communication system. The objective of these anticollision algorithms is to separate individual signals from one another. The anticollision algorithm is also known as multi-access algorithm. Multiple-access techniques are used in today's communication system for allowing multiple users to coexist in the same channel without having any interference or having minimum interference from

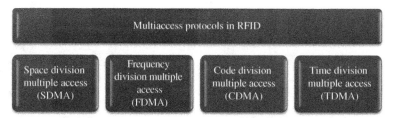

Figure 6.9 Taxonomy of anticollision protocols for RFID tags.

others. There are established anticollision algorithms that are being widely used in wireless communication systems. Some of them are used in conventional chipped RFID system. Most of these algorithms require some intelligence within the tag to establish a two-way communication between the reader and tags. This section provides a brief overview about the anticollision methods that are being used in conventional RFID system. Figure 6.9 shows taxonomy of RFID anticollision protocols present in modern RFID system [2]. Some of these methods can be also used in chipless RFID tags [9].

6.4.1 SDMA

SDMA refers the technique of reusing the channel into spatially separated zones [3]. The main objective of employing SDMA in an RFID system is to reduce the probability of tag collision in dense multi-tag environment. The interrogation zone is divided into multiple virtual channels. For SDMA, there are two options. One is to reduce the read range of a reader and by bringing multiple readers together as shown in Figure 6.10a. In this way, each reader will cover a particular spatial area. However, the collision may still occur if any tag falls within the overlapping area of two readers. Then the tag is interrogated by two readers simultaneously, which may lead to a reader collision. A better way to exploit the spatial distribution of tags for multiple tag reading is to use a phased array antenna [10] for the reader, and the directional beam is pointed directly to a tag. To address each tag within the interrogation zone, the reader scans the area with the directional antenna beams until it detects a tag. After reading the aimed tag, it will continue scanning for more tags in the zone. Therefore, various tags can be separated by utilizing their angular position variation in the interrogation zone of the reader as long as the angular distance between two tags is greater than the beamwidth of the antenna. As shown in Figure 6.10b, Tag 5 and Tag 6 have an angular separation of θ_t, which is greater than the 3 dB beamwidth θ_b of the antenna. Hence, they can be

(a)   (b)

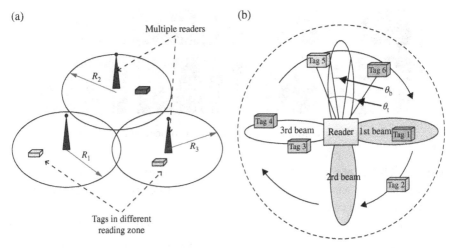

Figure 6.10 (a) Spatially separated reading zone with multiple readers. (b) Spatial separation of reading zone with phased array antenna.

separated by the directional antenna. But for Tag 3 and Tag 4, they fall within the same beam and collision occurs.

For near-field low-frequency RFID systems for LF and HF (below 850 MHz), the size of the array antenna is very big and not suitable for RFID reader. But for far-field high-frequency RFID systems (in GHz range), the array antenna size becomes smaller and can be installed with the RFID reader [11]. As shown in Figure 6.11, the phased array antenna consists of antenna elements, followed by a beamforming network, which is fed through an N-way power divider to feed an input signal to each antenna element. The elements are fed with relative phase shifts to create a narrow agile beam. For scanning with the array antenna, each element is fed with different amplitudes (currents) and phase shifts with voltage attenuators and variable phase shifters, respectively [12–14]. However, the beamforming network can either be an analog section with physical phase shifters and attenuators or a digital beamformer where the phase shifts and amplitude weighting is done in digital domain [15]. A smart antenna for employing SDMA in RFID has been reported in Ref. [16], which is operating between 860 and 960 MHz with a 10 dB gain and 40° elevation beam scanning capability. This low-cost, lightweight 3×2 array antenna can be implemented with a universal UHF RFID reader, which provides the reader with anticollision feature through SDMA.

A different approach to SDMA has also been reported in the literature where instead of a phased array antenna, multiple narrow-beam antennas have been used and they are being activated (switched) one

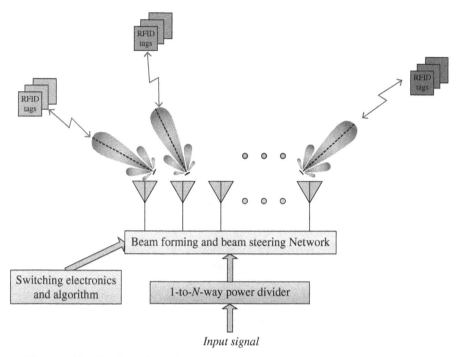

Figure 6.11 Configuration of phased array antenna for SDMA in RFID system.

by one by external switching network. In Ref. [17], SDMA method has been investigated using multiple antennas, each antenna is dedicated for a defined interrogation area. The antennas are activated one by one to identify tags in different spatial areas similar to Figure 6.10a. In addition, TDMA (ALOHA protocol) was considered for more discriminating multiple tags within the same spatial channel. In Ref. [18], six switched microstrip antennas in an array configuration have been designed for UHF RFID (915 MHz) tags. Each antenna has a 60° 3 dB beamwidth, and six of them are used to scan the 360° interrogation zone with 60° increment by activating one antenna at each step. The arrangement of antennas is shown in Figure 6.12. An electronic control section controls the switching among the antennas. When antenna-1 is activated, the communication takes place between the reader and tag set-1.

However, the drawback of SDMA is the high implementation cost of the antenna. The attenuators and phase shifters introduce noises in the system. It is unable to distinguish between two tags if they fall within the same directional beam. The use of SDMA is restricted to a few specialized applications [2]. A comprehensive review of the smart antennas for RFID can be obtained from Ref. [19].

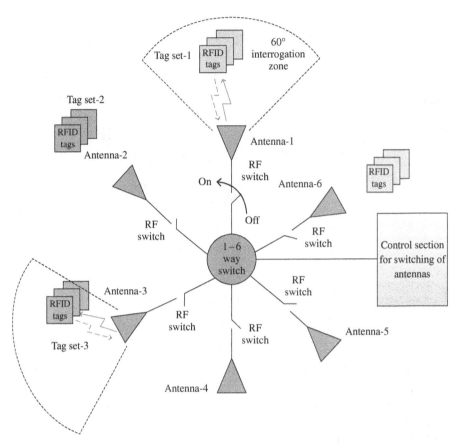

Figure 6.12 Geometrical arrangement of antennas for 360° scanning of interrogation zone with 60° step.

6.4.2 FDMA

FDMA refers to the technique of allocating several transmission channels of different carrier frequencies simultaneously available to each user in a communication system. For the implementation of this technique to RFID tag collision problem, the tags should be accomplished with the capability of freely adjusting transmission frequencies [23]. Tags have to select different carrier frequencies to modulate and send back their ID as shown in Figure 6.13a. The carrier frequencies should be different for all tags in the interrogation zone. Another FDMA approach has been proposed, and performance evaluation has been shown in Refs. [24, 25], which is termed as multicarrier backscattering. Instead of modifying the tags for FDMA, this method put the burdens on the interrogation side. Multiple continuous wave emitters (CWE)

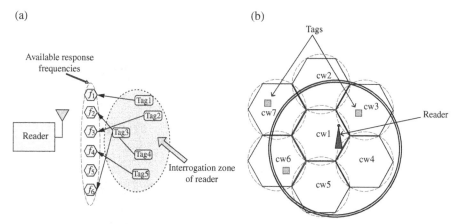

Figure 6.13 (a) In FDMA, the transponders have several frequency channels for transmitting their IDs. (b) Placement of readers in honeycomb style.

are arranged similar to cellular-based configuration as shown in Figure 6.13 to minimize cochannel interference. However, for receiving the response signals from interrogation zone, one reader unit is placed whose coverage is shown with solid circle. Now, each tag in a CWE's coverage zone uses the continuous wave emitted by the CWE to modulate the backscatter signal. As the CWEs emit different frequency signals, the reader is being able to separate multiple tag responses based on the carrier frequency.

Integrating the FDMA technique in RFID causes relatively high cost of reader as dedicated receiver for every reception channel. The tag also must possess the capability to select a frequency from available response frequencies, which makes the tag complex and costly. Hence, the use of this procedure is limited to very few applications.

6.4.3 CDMA

In traditional spread-spectrum (SS) CDMA systems, each user encodes its data using a spreading code, which is designed to be orthogonal or as close to orthogonal as possible to the other codes. This allows the successful decoding of data sent by two or more users simultaneously [25, 26]. Figure 6.14 shows the basic working principal of an RFID system with CDMA as a multiple-access scheme. However, for application of CDMA in RFID, the transponders must be capable of generating pseudorandom spreading code, creating spreading data, and modulating and reflecting the incoming RF signal with the spreading

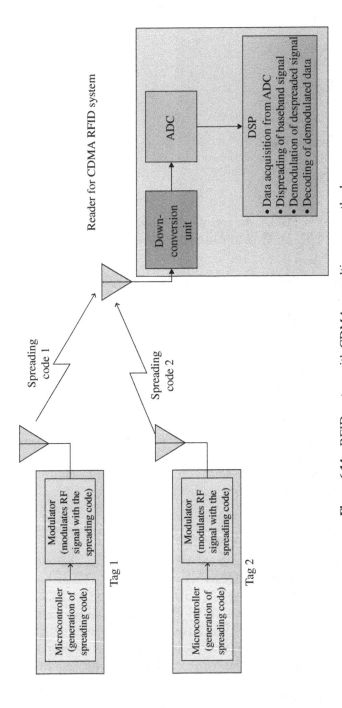

Figure 6.14 RFID system with CDMA as multi-access method.

data [23, 25]. In Figure 6.14, the transponders are replying simultaneously with a different spreading code. In Refs. [25–27], gold sequence has been used for spreading the tag IDs. Therefore, in the receiver section, the signal processing unit is capable of separating individual tag IDs by dispreading the combined backscattered signal with respective dispreading codes. The transponders reply to the reader simultaneously but with spreading their ID using unique spreading code. Thus, the received backscattered signal is a superimposed signal. Receiving this signal, the reader separates the tag responses by dispreading the received signal [3, 25, 26].

CDMA is ideal in many ways such as secure communication between RFID tags and reader and multiple tag identification. However, it adds quite a lot of complexity and computationally too intense for RFID tags. Therefore, it is not implemented in all types of RFID systems but in some specialized cases where secure identification together with multiple access is required such as authentication of secured and confidential documents.

6.4.4 Time Division Multiple Access: TDMA

TDMA is the largest group of anticollision algorithms used in RFID [2]. It is a technique to use the entire available channel capacity among the users chronologically. The reading time is divided into multiple time slots to be accommodated by the tags. This scheme is widely used in digital mobile radio systems.

Figure 6.15 shows a scheme of TDMA system. Different transponders respond in different time slots. In Figure 6.15, Tag 3 and Tag 4 respond in the same time slot 3 that causes a collision. There are various TDMA approaches for collision avoidance/multiple access in RFID tags. They can be primarily divided as tag driven/tag talks first (TTF) and reader driven/reader talks first (RTF).

6.4.4.1. *Tag Driven/TTF*   Tag-driven protocols function asynchronously [10, 11]; the data transfer is not controlled by the reader. The tags automatically send their IDs after being entered into the reading zone of the reader. It then waits for the reader to send acknowledgment. A positive acknowledgment (ACK) indicates successful reception of tag ID and a negative acknowledgment (NACK) means a collision. When two or more tags respond, partial or complete collision occurs. In supertag approach, which works on ALOHA-based anticollision protocol, tags continuously transmit their IDs at random intervals

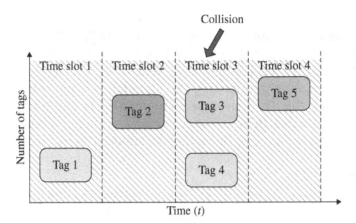

Figure 6.15   TDMA in RFID system.

until an acknowledgment is sent by the reader. The multiple access is achieved by muting or slowing down one tag that has been successfully identified. The number of tags in the interrogation zone is reduced after each successful tag response. Another variation involves muting of all tags in the interrogation zone except one being read. Then the tags are activated one by one. It ensures no collision among the tags. The protocol flowchart is shown in brief in Figure 6.16. The ALOHA algorithm is a probabilistic approach. Tag-Driven protocols are naturally slow. Hence, in most applications reader-driven protocols are being used.

6.4.4.2. Reader Driven/RTF   Reader-driven protocols are controlled by the reader as the master. These protocols are considered to be synchronous as all the tags in the interrogation zone are controlled and checked by the reader simultaneously [3]. An individual tag is first selected from a cluster of multiple tags in the reader's interrogation zone using certain algorithm, and communication with the tag is carried out, and tag ID is identified. Such protocols can be subdivided into polling, splitting protocols, and I-code. In polling methods, the reader has a list of all probable tag IDs, and it interrogates each ID one after another and if any tag ID matches with the interrogated one replies back to the reader. The flowchart is shown in Figure 6.17.

In splitting protocols, the colliding tags are subdivided into two subsets, and each subset is allowed different time slots to send the IDs. If the subset encounters collision, then they are further subdivided and

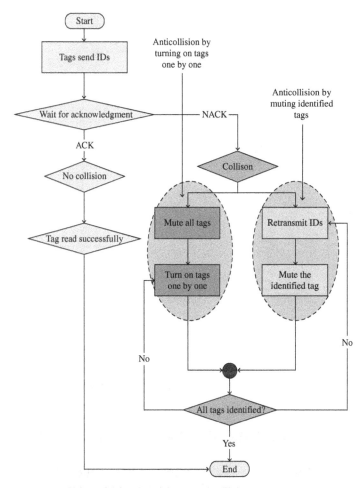

Figure 6.16   Tag-driven anticollision protocol.

the process goes on until all tag IDs are identified. The splitting protocols are further subdivided according to the methods of splitting in disjoint subsets. The subdivisions are tree splitting, query tree, binary search, and bit arbitration [2, 3, 20]. I-code is a stochastic process [20, 21, 22] using framed slotted ALOHA protocol. The reader starts with an estimation of frame size for identifying all tags. Tags randomly select a slot to send their IDs.

However, among all different types of anticollision protocols, TDMA is the one that is used in most of RFID applications. In some cases, TDMA with FDMA or with SDMA is used for more reliable communication between reader and multiple tags.

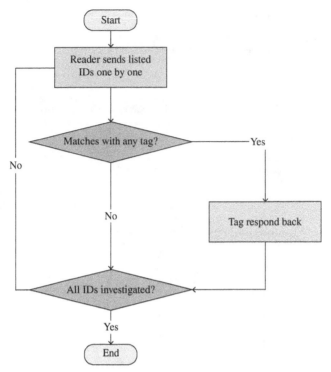

Figure 6.17   Flowchart of polling anticollision method.

6.5   ANTICOLLISION ALGORITHM FOR CHIPLESS RFID

Chipless RFID is a low-cost alternative of conventional RFID system, which is meant to provide all the functionalities of RFID but with a fraction of cost. Therefore, this technology is becoming popular day by day and attracting researchers to focus on various aspects of chipless RFID. However, like chipped tags, chipless tags also suffer from collision problem when multiple tags simultaneously respond back to the reader. It would be easier and straightforward solution if the available anticollision algorithms can be applied to the chipless RFID system with or without little modifications. However, this is not the state as chipless RFID system's working principal is way different from the conventional RFID system because of unique data encoding techniques. Chipless tags are printed passive structures. There is no onboard active element. IC-based tags can be activated and deactivated according to reader's query, whereas chipless tags are always active as long as they remain within the range of a reader antenna. The chipless tags are unable to sense any command from the reader. The tags are not suitable for complex signal processing within it. Thus, TDMA, FDMA, and CDMA techniques are not directly applicable to chipless RFID system. In some means, FDMA can be allocable for frequency signature-based chipless RFID tags if the tags are encoded in different frequency signatures and can be read in different times. The key anticollision method should be implemented in the reader side. However, among all the anticollision algorithms available, the SDMA technique is applicable for multi-access in chipless RFID. But SDMA causes high implementation cost, and when multiple tags fall within the beamwidth, it fails to separate them. Hence, special collision detection and resolution algorithms are required for chipless RFID system.

A few reported works are available on collision problems in chipless RFID systems. They are mainly focused with the SAW-based tags. They are compatible with TDR-based tags as almost similar operating principles are used for TDR-based tags. But till date, no reported literature has been found to address the problem of tag collision in frequency-domain chipless RFID tags [28–30]. The frequency-domain chipless tags have been reported to have up to 128 bits per tag, and they have a promising future for mass-level deployment in various fields for identification and tracking. A review based on available literature is discussed in the following. Therefore, to improve the reliability of reading and to provide it with multi-access capability, a systematic research on multi-tag scenario is a promising area to work on. The collision avoidance algorithms used for SAW tags are shown in Figure 6.18.

6.5.1 Linear Block Coding

Linear block coding was proposed as a separation method for simultaneous interrogation of multiple SAW tags [31, 32]. It has been proposed as an anticollision method where digital beamforming or other (SDMA) technique fails to separate the tags. In such a scenario, the adaptability for multiple tag reading is needed to be incorporated in the signal processing methods. In SAW tags, the ID is decoded estimating the round-trip delay time (RTDT) of the response echo of the tag. Using linear block coding technique, the code structure of the tag is modified to guarantee unique superposition patterns. Figure 6.19 shows a SAW tag with interrogation unit. Multiple reflections occur from the tag, which is used as tag ID. Figure 6.20 shows the modified structure of the tag for block coding. The coding information is extracted from the tag's impulse response. The position of each reflector is a part of linear block

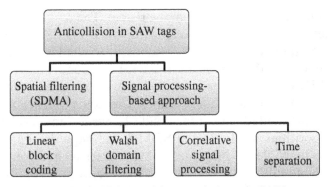

Figure 6.18   Collision avoidance techniques in SAW tags.

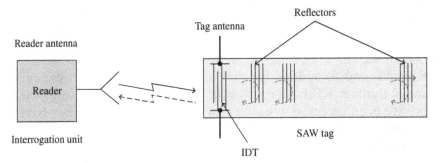

Figure 6.19   Interrogation unit with SAW tag [28].

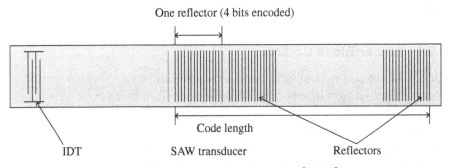

Figure 6.20   Layout of tag for coding [28, 29].

code. Within one group, only one reflector is allowed per tag. The coding information is extracted by estimating the combined RTDT from the impulse response of the SAW tag. When the interference of two tags is destructive, the corresponding peak amplitude decreases, which causes the corresponding RTDT to be treated as invalid. The valid RTDTs are converted to binary values and validated using block code.

The block coding scheme is used to detect and correct bit error that may occur due to interference of other tags or multipath effect. It is to be noted that a certain amount of bit error can be detected and corrected. When too many bit errors occur, then the tags need to read out again. Here the authors have concentrated for separation of at least two tags. Using Block coding as a coding scheme adds some flexibility and reliability in the decoding procedure. Missing information due to multipath or other effects can be recovered. Error correction and also separation of two tag response becomes possible in some cases. For resolving two tags the coding scheme works well but becomes critical and non-linear for more than two tags.

6.5.2   Correlative Signal Processing-Based Approach

Correlative signal processing for identifying a particular response signal buried in a collided or distorted signal has been discussed in Refs. [33, 34] for SAW sensor tags. Individual sensors are coded differently, but they are not completely time orthogonal. All the sensors are interrogated with the same RF burst. The receiver peaks the superposition responses from multiple sensors. The receiver uses correlative signal processing for extracting and separating individual sensor response [35]. The composite response is correlated with a replica of sensor response, which is to be separated or extracted from the collided signal. When the signal matches best, the cross-correlation function has the highest correlation peak. This way, individual sensor signal is separated from the collided response. The block diagram of the whole method is shown in Figure 6.21. The maxima of each correlation output are detected by the maxima detection unit. The final separation based on the maxima of CCF is processed in the postprocessing unit, which also controls the whole method.

6.5.3   Walsh-Domain Matched Filtering

Another approach for resolving collision problems in SAW tags has been proposed in Ref. [36] where a matched filtering technique has been used in Walsh domain for separating collided responses from multiple tags. It has been reported that though the collided signals cannot be distinguished in time domain, they pose special features in Walsh domain. Hence, Walsh-domain matched filtering has been proposed here. A database is maintained that contains all possible tag response signals. The sum of tag response signal is match filtered in Walsh domain with a replica of single tag response that is to be extracted from the

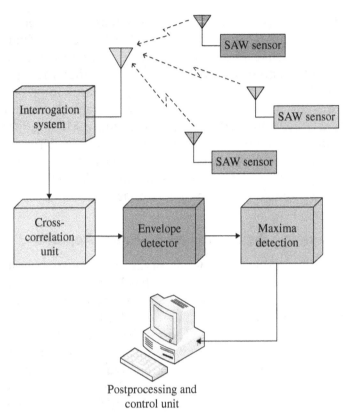

Figure 6.21 Block diagram for correlative signal processing multi-access method in SAW sensors.

collided signal. A reasonable threshold is used to improve the operating speed. The algorithms have been verified through a prototype implementation and reading multiple tags simultaneously. It shows the distinguishable property of multiple SAW tags even if the tags are read synchronously and reflector pulses overlap. SAW tags are purely passive devices, and their response cannot be controlled or stopped according to reader's instruction like traditional chipped RF tags. Hence, special separation algorithms are required to separate their collided response, and this method shows acceptable result in prototype testing (Fig. 6.22).

6.5.4 Spatial Focusing (SDMA)

Spatial focusing using an agile narrow-beam antenna is an attractive solution for avoiding collision among chipless RFID tags. This is termed as SDMA, which has already been described in Section 6.4.1 elaborately.

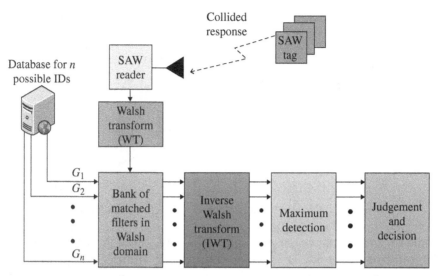

Figure 6.22 Walsh-domain filtering for multi-tag decoding in multiple SAW tag environment.

The section described the SDMA approaches used for conventional chipped RFID system. As the chipless tags cannot employ smart signal processing techniques within the tag, SDMA seems to be the attractive solution for tag collision in chipless RFID system. Therefore, SDMA for chipless RFID is a research area of significant interest [10, 37]. A lot of extensive research is going on for beamforming and beam steering with smart antenna both in analog and digital domain beamforming. However, little reported work is found that fully dedicates the SDMA approach with smart antenna for chipless RFID readers. In Refs. [6, 34], SDMA approach with an array antenna has been reported for multiple SAW tag identification where a 2.54 GHz focused beam antenna was used with 1.5 m focal length. However, interference can occur if two tags are simultaneously illuminated by the narrow-beam antenna. Therefore, using a single anticollision method for chipless RFID system does not offer the completeness in reliability and multiple access. So SDMA need to be used with other multiple-access methods (Fig. 6.23).

In Ref. [10], a phased array antenna approach for multiple frequency-domain chipless RFID tag identification has been proposed. The chipless RFID system developed by Monash University Antenna and RFID research group [28–30] has already been discussed in the previous chapters. The chipless RFID system operates through the UWB band (3–10 GHz). Being a wideband system, it is challenging to obtain a

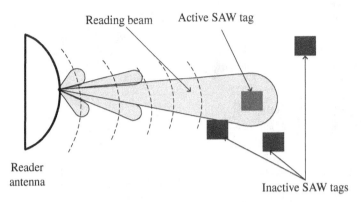

Figure 6.23   A focused beam antenna for selectively reading tags [6].

uniform performance from a phased array antenna with electronic beamformer network as the electronic phase shifters, power dividers, etc. are inherently narrowband [10]. Therefore, optical beamforming has been considered as a viable option for a smart chipless RFID reader with multi-access protocol. A hybrid optical module was used as a beamforming network with 8-element array antenna. The optical phase shifters have the ability to form multiple beams as well as steering of beam in 3D plane.

The phased array antenna for the smart chipless RFID reader not only provides multiple tag reading facility but also improves the performance of the system by discarding undesired signals and interferences by pointing the beam directly to the desired tag. Therefore, it is an excellent candidate as reader antenna for smart chipless RFID reader.

6.5.5   Other Anticollision/Multi-Access Methods

There are several other anticollision proposals that have been found in the literature [37] but have not been found any prototyping application till date. They include signal strength-based method [37], time separation [37], and signal subtraction [37]. Signal strength-based capture effect method separates tag response signals based on the strength of the signals to the reader antenna [38]. The closer one will have more strong response signal than others. Such a method is suitable for application for systematic arrangement of tagged items such as boxes on a conveyer belt. In time separation-based anticollision algorithm, the tags are encoded with inherent and unique delays so that two tags do not reply simultaneously [37]. However, such anticollision is suited for a situation where a hierarchy of tag types is needed. This anticollision method is restricted because of limited availability of nonoverlapping time segments in practical scenario. In signal subtraction-based method, the strongest tag signal is decoded using matched filtering technique [37]. Then an ideal replica of the signals is mathematically generated and subtracted from the composite signal. In this way, the interfering signals are removed, and tag response signals are separated from the mixed response signal.

6.6   COLLISION DETECTION AND MULTIPLE ACCESS FOR CHIPLESS RFID SYSTEM

It is evident from earlier discussion that chipless RFID system is different from the conventional RFID systems in many ways. Therefore, the available anticollision and multiple-access algorithms used in wireless communications and conventional RFID system are not straightaway applicable to chipless RFID systems. In conventional RFID, the tag contains active elements (ASIC, battery), and hence, the sleep and wake-up states of a tag can be controlled remotely from the reader. The tag circuit can be modified for sensing collision in the wireless channel, and it is capable of establishing a two-way communication with the reader by sensing the different query signal from the reader. Therefore, various ways of establishing a multiple-access algorithm are available in conventional RFID system. On the contrary, the chipless RFID system has much more similarity with RADAR system from the tag's perspective as the tags do not possess any intelligence. Just like the RADAR target, they backscatter the incident signal to various directions, and the reader antenna captures a portion of the backscattered signal and analyzes the spectrum of the signal to find the resonance information, which denotes the tag ID. Therefore, the multiple tag detection and collision avoidance challenges need to be addressed from signal processing perspective together with the smart antenna for employing SDMA. Thus, the difference in geometrical distribution of tags can be exploited as a distinguishing criterion for multiple tags. Certainly, this situation will be application specific.

For the signal processing approach, two methods based on time difference of arrival (TDOA) of response signals from different tags are investigated. The first method is based on joint time–frequency analysis (JTFA) using fractional Fourier transform (FRFT) [39] where the collided response signals from multiple tags (for simplicity, simultaneous responses from two tags have been considered) are analyzed in time–frequency plane. When the tags have different geometrical distances from the tags, they will have measurable delay (TDOA) between their response times. But this delay is very small, and the responses overlap in time domain. However, as the tags are backscattering the same interrogation signals, they will have identical frequency contents. So neither time-domain windowing nor frequency-domain filtering is capable of separating them. The fractional Fourier transform (FrFT) converts their response signals to the optimum fractional domain, and because of the delay between them, the two response signals will be concentrated at two different regions with some temporal overlaps. The windowing in fractional domain is capable of separating them from each other. Afterward, from the spectrum of the separated signals, the individual tags' IDs are extracted. A simplified block diagram of the proposed method is illustrated in Figure 6.24. The details of the method will be presented elaborately in Chapter 7.

The second method is similar to the well-known FMCW-RADAR technique. A linear frequency-modulated (LFM) signal is sent out by

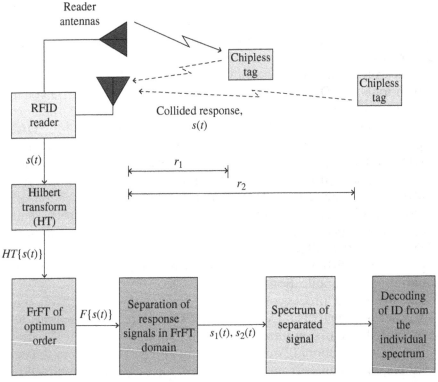

Figure 6.24 Time–frequency analysis through FrFT for multiple tag identification in chipless RFID system.

the reader's interrogation antenna to the reading zone. The tags back-scatter the signal after modulating with their tag IDs. The received signal is downconverted to the intermediate frequency (IF) signal through a mixer. However, when only one tag is responding from the reading zone, the IF signal ideally contains only a single frequency signal corresponding to the tag response. This signal is analyzed further through Hilbert transform (HT) as proposed in [40] to identify the tag ID. When more than one tag respond back with measurable delay among individual tag responses, the IF signal consists of multiple beat frequency signals, each related to a particular tag. Therefore, by analyzing the spectrum of IF signal, the number of beat frequency signals is estimated, which in turn denotes the number of collided tags. If the tags are well separated in their positions, their beat frequencies will also be well separated in frequency domain after the mixer. They will fall in different frequency bins. Therefore, it will be possible to separate them by filtering in frequency domain. For filtering out each signal, a bank of narrowband filter or a filter whose frequency can be continuously adjusted with narrow detection bandwidth [41] is used. The overall system block diagram and processing steps are shown in brief in Figure 6.25. The details will be found in Chapter 7.

However, as already discussed, for RFID system, in most cases, a single anticollision algorithm is not sufficient for multi-tag environment. Therefore, in most of the applications, a combination of anticollision algorithms is employed to have robust and reliable communication between RFID reader and tags. The ultimate goal is to use the signal separation methods together with SDMA with a phased array antenna

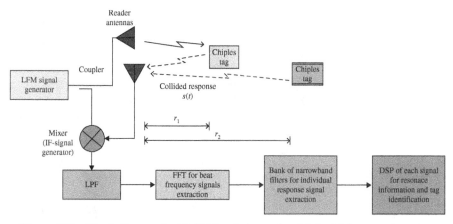

Figure 6.25 Block diagram for FMCW technique of multiple chipless tag detection.

[10, 14] so that when more than one tag are illuminated by the narrow-beam array antenna, the reader is still capable of detecting collision. More insight about the proposed methods is provided in next two chapters.

6.7 INTRODUCING BLOCK CODING IN CHIPLESS RFID

The data integrity and authenticity of an RFID system are challenged by various factors. While the wireless communication takes place between the reader and RFID tags in the interrogation zone through air interface, due to noise and other interferences, the probability of successful identification decreases. The identification error can occur due to many factors as noise, data collision from nearby tags, and reduction in return signal level due to increased distance as shown in Figure 6.26. Now, the question is what is meant by "identification error." The RFID reader decodes the ID of a tag from the received signal. Two possible bit values are "0" and "1." As already described, in a frequency-domain chipless RFID system, the presence and absence of resonance is decoded as "1" and "0," respectively. However, the resonances are identified by comparing the signal level at particular frequencies compared to nearby frequencies [42]. Due to transmission error, a wrong decoding

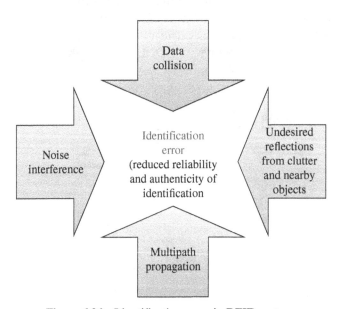

Figure 6.26 Identification error in RFID system.

may occur, which may alter any bit in the decoded tag ID. For example, instead of identifying ID "1010," due to transmission error, the reader is decoding the ID as "1011." This is termed as identification error. To address the problem, current chipped RFID systems apply a cyclic redundancy check in some applications to improve the identification performance [1].

In chipless RFID system, the data bits are encoded into 1:1 correspondence, without any parity or error check bits [28–30] as the main aim is to incorporate as much data bits as possible within a small footprint. The reported literature [43, 25] shows that between 3.1 and 7 GHz 35 bits (over 1.3 billion ID combinations) has been achieved within a small footprint as to fit in any Australian banknote for retransmission-type chipless tags. As reported in [44], using slot-loaded monopole antenna, 128 bits has been achieved so far within a small dimension within the UWB (3.1–10.7 GHz). However, neither of the reported works has considered the problem of data loss and data integrity during transmission. As mentioned before, the high data capacity, low-cost chipless RFID tags with all the advantages of RFID system are becoming an excellent alternative for costly chip-based RFID tags. With the expansion of application fields, the requirement of identification reliability and authenticity are coming into being. Therefore, the aim of this chapter is to provide some insight about the probable approaches for reliability improvement for identification in chipless RFID systems. Though the main burden of reliability and data integrity lies on the reader part of the chipless RFID systems, within the limited and small room available within the tag, this chapter aims to modify the encoding method of the tag so that together with the reader, the tag can also participate for reliability improvement. The block coding is introduced for designing the multiresonator section of chipless RFID tag. The effect and usability of block coding has been described from two directions. One is for collision detection in multi-tag environment, and the second one is for error detection and correction in received data bits for reliability improvement.

6.7.1 Coding

In block coding, the binary information sequence is segmented into message block of fixed length; each of the blocks consists of m information bits and k parity or error check bits. Therefore, instead of directly encoding the IDs, code words are generated with parity/check bits, and the code words are used as tag IDs. If we consider 4-bit tags, without block coding, 4 bits are encoded within the tag. However, if we want to include three check bits for bit error detection, then for a 4-bit tag ID, we have to use seven bits, three for error check and four for tag ID. A generator matrix is used to generate code words from individual IDs. Then the chipless tags are designed to respond according to the code words for identification. In the receiver side, the same generator matrix is used to decode the tag ID from the code words. Below a $(7, 4)$ hamming code words are generated for 4-bit chipless tags with three check bits. The code words are generated from the tag IDs using generator matrix, **G**, by [45]

$$\mathbf{c} = \mathbf{m}.\mathbf{G} \tag{6.1}$$

where $\mathbf{c}$ = codeword, $\mathbf{m}$ = original tag ID, and $\mathbf{G}$ = generator matrix.

The generator matrix depends on the arrangement of the bits in the code word. The set of code words generated through the aforementioned method is called *block codes*. Now, for 4-bit tags, 16 different tag IDs are possible, and we need four resonators for designing the tags as shown in Figure 6.27a. However, for including check bits for collision and error detection, instead of sixteen 4-bit IDs, there will be 16 different

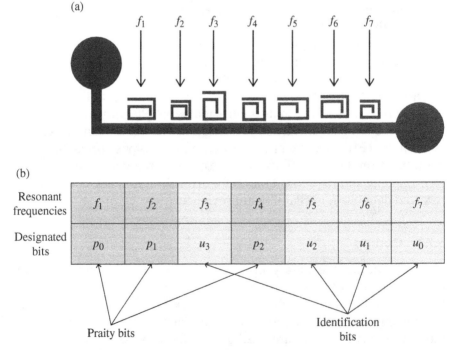

Figure 6.27   (a) Layout of 7-bit chipless tag and (b) arrangement of ID bits and check bits in code word [45].

7-bit code words. Therefore, seven resonators resonating individually (without affecting the nearby resonators) at seven different frequencies as $f_1, f_2, ..., f_7$ are required. The identification bits and check bits are arranged as shown in Figure 6.27b.

It is evident from Figure 6.27 that the first, second, and fourth resonances are used for check bits and other resonances are used for tag ID. If a total of 128 bits are available, they can be grouped according to the method described here. Each block will have seven bits allocated among which four bits are used for tag ID and three bits as check bits. A total of 18 blocks will be there, and 72 bits are available for tag ID, which provides 2^{72} numbers of tag IDs. However, here, we only consider one block of data with four data bits and three check bits. So it resembles a (7, 4) block code. For (7, 4) block code, the entire code words generated for 4-bit IDs are shown in Table 6.1.

From the previous table, it is evident that when the tag ID is "0011," the corresponding code word will be "1000011." The response from the tag has resonances at $f_1, f_6,$ and f_7.

6.7.2 Block Coding for Collision Detection

Figure 6.28 shows a smart library shelf with RFID tagged books. However, as already discussed, the anticollision methods and algorithms in RFID systems are not universal for all applications rather it is application specific. Different applications require different anticollision methods to be employed for multi-access and collision avoidance. As shown in Figure 6.28, for the first antenna position, there is no collision, and hence, reliable and error-free identification can be done. But in the second antenna position, at least two books are illuminated by the narrow beam of the reader antenna, which gives rise to collision, which in turn causes error in decoded ID and decreased reliability of the identification system. Figure 6.29 shows the decoding procedure of tag IDs in current chipless RFID readers. The spectrum of the response from tag is analyzed and based on the presence and absence of resonance a "1" and "0" is decoded for individual bits. However, through this decoding and identification method, the chipless RFID reader is neither capable of collision detection nor decoding error detection. Therefore, intermediate steps are required for collision detection and error correction.

Linear block coding (LBC) has already been introduced in the previous section, and in the literature, it has also been found to be used for simultaneous reading of two SAW tags and error detection. Here, our aim is to initially validate the idea of introducing block code in

TABLE 6.1 The (7,4) Coding for RFID Tag Identification

	Code Word and Resonance Corresponding to Tag ID												
	1st Bit/Resonance		2nd Bit/Resonance		3rd Bit/Resonance		4th Bit/Resonance		5th Bit/Resonance		6th Bit/Resonance		7th Bit/Resonance	
Tag ID	p_0	f_1	p_1	f_2	u_3	f_3	p_2	f_4	u_2	f_5	u_1	f_6	u_0	f_7
0000	0	×	0	×	0	×	0	×	0	×	0	×	0	×
0001	1	√	1	√	0	×	1	√	0	×	0	×	1	√
0010	0	×	1	√	0	×	1	√	0	×	1	√	0	×
0011	1	√	0	×	0	×	0	×	0	×	1	√	1	√
0100	1	√	0	×	0	×	1	√	1	√	0	×	0	×
0101	0	×	1	√	0	×	0	×	1	√	0	×	1	√
0110	1	√	1	√	0	×	0	×	1	√	1	√	0	×
0111	0	×	0	×	0	×	1	√	1	√	1	√	1	√
1000	1	√	1	√	1	√	0	×	0	×	0	×	0	×
1001	0	×	0	×	1	√	1	√	0	×	0	×	1	√
1010	1	√	0	×	1	√	1	√	0	×	1	√	0	×
1011	0	×	1	√	1	√	0	×	0	×	1	√	1	√
1100	0	×	1	√	1	√	1	√	1	√	0	×	0	×
1101	1	√	0	×	1	√	0	×	1	√	0	×	1	√
1110	0	×	0	×	1	√	0	×	1	√	1	√	0	×
1111	1	√	1	√	1	√	1	√	1	√	1	√	1	√

√, denotes presence of resonance; ×, denotes absence of resonance.

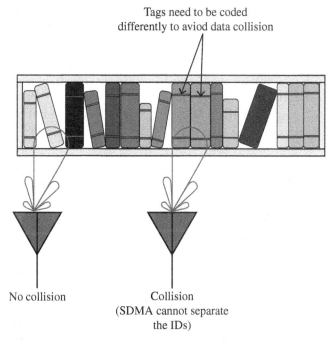

Figure 6.28 Collision of tagged items in smart library, requirement of coding.

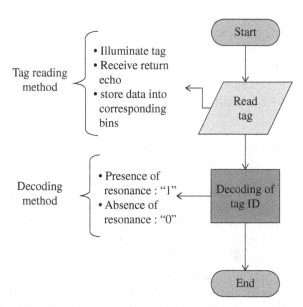

Figure 6.29 Flowchart for chipless RFID tag ID decoding in reader.

chipless RFID tags for collision and error detection. By introducing block coding, the resonator section of chipless RFID tag is modified according to Table 15.1 by turning on and off particular resonators based on the code word corresponding to a particular tag ID. When simultaneous responses are received by the reader from multiple tags, they cause decoding errors by alteration of different bit values. For example, two tags with ID "0100" and "1010" are simultaneously sending back their individual response signals to the reader. The reader will consider having three resonances in the received signal and decoding the tag ID as "1110" as shown in Figure 6.30a. However, there is no way for the reader to either detect a collision or validate the tag ID.

Now, as the block coding is used, instead of "0100" and "1010," the tags are having code words as "1001100" and "1011010," respectively, as shown in Figure 6.30b. Now, the decoded ID from the collided signal can be verified based on block coding. The flowchart in Figure 6.31 explains the collision detection steps through code validation. The binary values of the code word are decoded from the resonance of the received signal. The binary values of the decoded code words are validated on the basis of block code by three steps. In the first step, the XOR operation is performed on the odd-bit values (1st, 3rd, 5th, and 7th). If the ID is correctly decoded, it should be "0." But here, it will have a value of "1," and the reader concludes that data collision has been occurred in the received data bits. If this value is zero, then next step for validation is to take the XOR result of the 2nd, 3rd, 6th, and 7th bit values. If it turns to be "1," then a collision is detected; otherwise, the next step of validation is performed, which is the XOR value of the 4th, 5th, 6th, and 7th bit values. If this is also "0," then it is concluded that the tag ID is correctly decoded. Otherwise, it is concluded that data collision are happening from multiple tags.

6.7.3 Block Coding for Improving Data Integrity

This section describes the application of block coding and check bits for bit error detection in decoded tag ID for improving the data integrity and identification reliability. As already discussed, the alteration of bit values may arise during the identification, which leads to wrong identification. This degrades the reliability and authenticity of the chipless RFID system. The block coding method used here is capable of single bit error detection and correction.

The steps for error detection and correction are shown in the flowchart of Figure 6.32. For example, the tag is having the code word C_4. So the

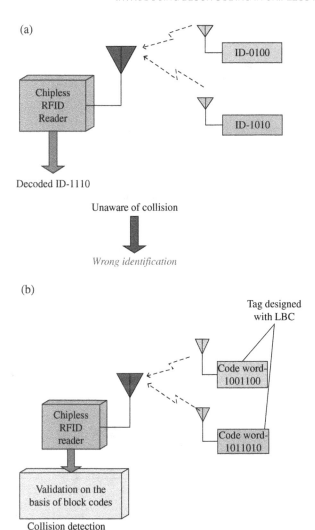

Figure 6.30 (a) Collision of two tags, IDs in 1:1 correspondence; (b) Collision of two tags where resonator section is designed with LBC for each tag.

decoded bit stream should be 1000011. But due to error, the decoded bit stream is 1100011. Now, after following the validation procedure, we have 010, which show that bit error is in the 2nd bit position. As only two states are possible, either 0 or 1 (depending on the presence or absence of a dip), it is clearly understandable that the bit we received for the 2nd bit has been altered. So the code word can be corrected as 1000011. However, the code used here is capable of detecting and correcting single bit error. If more than one erroneous bit is decoded, then this

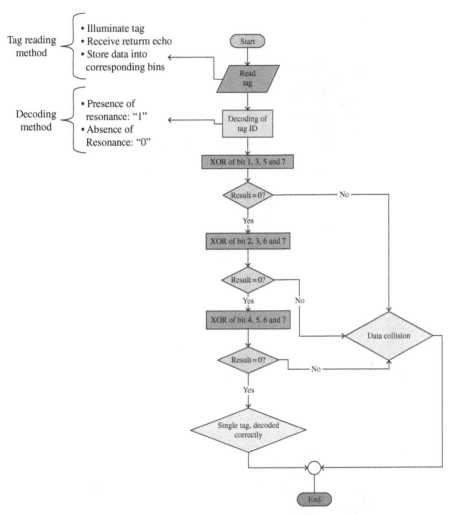

Figure 6.31   Flowchart for collision detection with block coding.

method is unable to detect and correct that. For this purpose, more check bits need to be used for each of the identification data block.

6.7.4   Advantages and Challenges of Block Coding

The advantage of this coding scheme is that it will provide us with a robust decoding method where we will be able to detect the presence of bit error and also to correct the erroneous bit to some extent in case of single tag reading. For multiple tags, this can be used to identify whether a collision is there or not. The challenge is

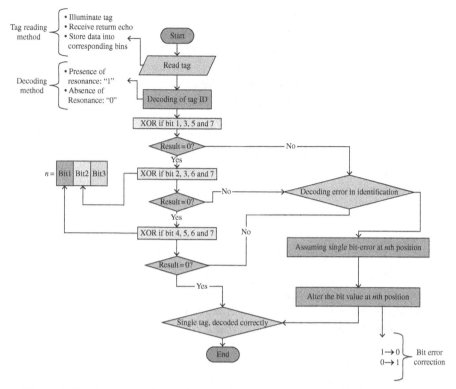

Figure 6.32 Error detection and correction method for single bit decoding error.

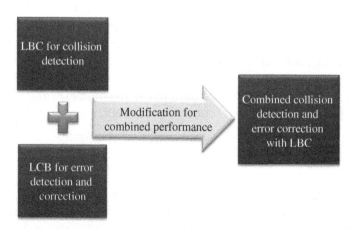

Figure 6.33 Probable modification of LBC for collision and error detection simultaneously.

to optimize the coding method for single tag reading error detection and multiple tag detection simultaneously. This is a potential research area as to modify the coding method in such a way so that it can be used for both purposes simultaneously as shown in Figure 6.33. However, the reader of the chipless RFID system needs to be incorporated with intelligent signal processing algorithms together with coding in tags for more reliability and data integrity in the chipless RFID system. The next two chapters give some insight about the intelligent signal processing methods for reader for collision handling and identification improvement in chipless RFID system.

6.8 CONCLUSION

Chipless RFID is a breakthrough in the field of RFID and Auto-ID. The potential application areas and acceptability of chipless RFID are emerging day by day. This comes with additional challenges like multiple access, collision detection, and data integrity for providing more reliable, authentic, and robust identification with chipless RFID systems over other existing systems. However, this chapter provided a comprehensive study about the established anticollision methods for different RFID systems. An inclusive review about collision avoidance and error check methods for chipped and chipless RFID systems has been presented. It is evident from the review that straightaway application of available anticollision methods is not a feasible solution for chipless RFID system. However, the potential of chipless RFID system for low-cost item tagging, tracking, and Auto-ID is beyond question. Therefore, intensive research and investigations are required for multiple access and data integrity in chipless RFID system. Smart signal processing approaches and linear block coding have been explored for addressing the challenges. This chapter will provide an intuitive and complete understanding about collision and error correction to the reader.

REFERENCES

1. Y. Lei, H. Jinsong, Q. Yong, W. Cheng, L. Yunhao, C. Ying, and Z. Xiao, "*Revisting Tag Collision Problem in RFID Systems*," in *39th International Conference on Parallel Processing (ICPP), 2010*, San Diego, CA, September 13–16, 2010, pp. 178–187.

2. K. Finkenzeller, *RFID Handbook: Fundamentals and Applications in Contactless Smart Cards and Identification*, John Wiley & Sons, Inc., New York, 2003.

3. T. Zhijun and H. Yigang, "*Research of Multi-Access and Anti-Collision Protocols in RFID Systems*," in *IEEE International Workshop on Anti-counterfeiting, Security, Identification, 2007*, Xiamen, Fujian, April 16–18, 2007, pp. 377–380.

4. Available: *RFID Forecasts, Players and Opportunities 2011–2021* http://www.idtechex.com/research/reports/rfid_forecasts_players_and_opportunities_2011_2021_000250.asp (Date accessed: June 16, 2012).

5. http://www.hidglobal.com/products/rfid-tags (Date accessed: June 16, 2012).

6. http://www.rfid-weblog.com/50226711/rfid_has_18_advantages_over_barcodes.php (Date accessed: June 14, 2012).

7. http://www.essenrfid.com/products-high-performance/tag_Parka.htm (Date accessed: June 12, 2012).

8. Y. Jiexiao, L. Kaihua, and Y. Ge, "*A Novel RFID Anti-Collision Algorithm Based on SDMA*," in *4th International Conference on Wireless Communications, Networking and Mobile Computing, 2008. WiCOM '08*, Dalian, October 12–14, 2008, pp. 1–4.

9. C. Hartmann, P. Hartmann, P. Brown, J. Bellamy, L. Claiborne, and W. Bonner, "*Anti-Collision Methods for Global SAW RFID Tag Systems*," in *2004 IEEE Ultrasonics Symposium (Vol. 2)*, Montréal, Canada, August 23–27, 2004, pp. 805–808.

10. N.C. Karmakar (Editor), *Smart Antennas for RFID Systems*, Wiley Microwave and Optical Engineering Series, John Wiley & Sons, Inc., Hoboken, NJ, 2010.

11. A. Baki, N.C. Karmakar, U. Bandara, and E. Amin, "Beam forming algorithm with different power distribution for RFID reader," in *Chipless and Conventional Radio Frequency Identification: Systems for Ubiquitous Tagging*, editor N.C. Karmakar, IGI Global, Hoboken, pp. 64–95, 2012.

12. N. C. Karmakar, P. Zakavi, and M. Kumbukage, "FPGA controlled phased array antenna development for UHF RFID reader," in *Handbook of Smart Antennas for RFID Systems*, Wiley Microwave and Optical Engineering Series, John Wiley & Sons, Inc., Hoboken, NJ, pp. 211–242, 2010.

13. N.C. Karmakar, P. Zakavi, and M. Kumbukage, "*Development of a Phased Array Antenna for Universal UHF RFID Reader*," *Digest 2010 IEEE AP-S International Symposium on Antennas and Propagation and 2010 USNC/CNC/URSI Meeting* in Toronto, ON, July 11–17, 2010, (cd-rom).

14. N. C. Karmakar, "Smart antennas for automatic radio frequency identification readers," in *Handbook on Advancements in Smart Antenna*

Technologies for Wireless Networks, editors C. Sun, J. Cheng, and T. Ohira, Information Science Publishing, Hershey, pp. 449–472, 2008.

15. J. Litva and T. K. Lo, *Digital Beamforming in Wireless Communications*, Artech House, Norwood, MA, 1996.

16. N. C. Karmakar, S. M. Roy, and M. S. Ikram, *"Development of Smart Antenna for RFID Reader,"* in *IEEE International Conference on RFID, 2008*, Las Vegas, NV, April 16–17, 2008, pp. 65–73.

17. H. Abderrazak, B. Slaheddine, and B. Ridha, *"A Transponder Anti-Collision Algorithm Based on a Multi-Antenna RFID Reader,"* in *Information and Communication Technologies, 2006. ICTTA '06. 2nd*, Damascus, April 24–28, 2006, pp. 2684–2688.

18. D. J. Kim, S. H. Kim, Y. K. Kim, H. Lim, and J. H. Jang, *"Switched Microstrip Array Antenna for RFID System,"* in *Proceedings of the 38th European Microwave Conference, 2008. EuMC 2008*, Amsterdam, October 27–31, 2008, pp. 1254–1257.

19. N.C. Karmakar and P. Zakavi, *"Compact Phase Shifter for UHF RFID Applications,"* *Digest 2010 IEEE AP-S International Symposium on Antennas and Propagation and 2010 USNC/CNC/URSI Meeting* in Toronto, ON, July 11–17, 2010, (cd-rom).

20. D.-H. Shih, P.-L. Sun, D. C. Yen, and S.-M. Huang, "Taxonomy and survey of RFID anti-collision protocols," *Computer Communications*, vol. 29, pp. 2150–2166, 2006.

21. L. Hsin-Chin and C. Jhen-Peng, *"Performance Analysis of Multi-Carrier RFID Systems,"* in *International Symposium on Performance Evaluation of Computer & Telecommunication Systems, 2009. SPECTS 2009*, Istanbul, July 13–16, 2009, pp. 112–116.

22. H. Liu and Y. Chen, "A frequency diverse Gen2 RFID system with isolated continuous wave emitters," *Journal of Networks*, vol. 2, pp. 54–60, 2007.

23. C. Mutti and C. Floerkemeier, *"CDMA-Based RFID Systems in Dense Scenarios: Concepts and Challenges,"* in *IEEE International Conference on RFID, 2008*, Las Vegas, NV, April 16–17, 2008, pp. 215–222.

24. W. Lih-Chyau, C. Yen-Ju, H. Chi-Hsiang, and K. Wen-Chung, *"Zero-Collision RFID Tags Identification Based on CDMA,"* in *Fifth International Conference on Information Assurance and Security, 2009. IAS '09*, Xi'an, August 18–20, 2009, pp. 513–516.

25. A. Loeffler, "Using CDMA as anti-collision method for RFID: research and applications," in *Current Trends and Challenges in RFID*, editor C. Turcu, ISBN: 978-953-307-356-9, InTech.

26. A. Loeffler, F. Schuh, and H. Gerhaeuser, *"Realization of a CDMA-Based RFID System Using a Semi-Active UHF Transponder,"* in *6th International Conference on Wireless and Mobile Communications (ICWMC), 2010*, Valencia, September 20–25, 2010, pp. 5–10.

27. D. K. Klair, C. Kwan-Wu, and R. Raad, "A survey and tutorial of RFID anti-collision protocols," *IEEE Communications Surveys and Tutorials*, vol. 12, pp. 400–421, 2010.

28. S. Preradovic and N. C. Karmakar, "*Design of Fully Printable Planar Chipless RFID Transponder with 35-bit Data Capacity*," in *European Microwave Conference, 2009. EuMC 2009*, Rome, September 29–October 1, 2009, pp. 013–016.

29. I. Balbin, "*Multi-Bit Fully Printable Chipless Radio Frequency Identification Transponders*," Ph.D. Thesis, Electrical and Computer Science Engineering, Monash University, Melbourne (2010).

30. I. Balbin and N. C. Karmakar, "Phase-encoded chipless RFID transponder for large-Scale low-cost applications," *IEEE Microwave and Wireless Components Letters*, vol. 19, pp. 509–511, 2009.

31. M. Brandl, S. Schuster, S. Scheiblhofer, and A. Stelzer, "*A New Anti-Collision Method for SAW Tags Using Linear Block Codes*," in *2008 IEEE International Frequency Control Symposium*, Honolulu, HI, May 19–21, 2008, pp. 284–289.

32. G. Bruckner and R. Fachberger, "*SAW ID Tag for Industrial Application with Large Data Capacity and Anticollision Capability*," in *IEEE Ultrasonics Symposium, 2008. IUS 2008*, Beijing, November 2–5, 2008, pp. 300–303.

33. G. Ostermayer, A. Pohl, R. Steindl, and F. Seifert, "*SAW Sensors and Correlative Signal Processing-a Method Providing Multiple Access Capability*," in *1998 IEEE 5th International Symposium on Spread Spectrum Techniques and Applications: Proceedings (Vol. 3)*, Sun City, September 2–4, 1998, pp. 902–906.

34. G. Ostermayer, "Correlative signal processing in wireless SAW sensor applications to provide multiple-access capability," *IEEE Transactions on Microwave Theory and Techniques*, vol. 49, pp. 809–816, 2001.

35. G. Ostermayer, A. Pohl, C. Hausleitner, L. Reindl, and F. Seifert, "*CDMA for Wireless SAW Sensor Applications*," in *IEEE 4th International Symposium on Spread Spectrum Techniques and Applications Proceedings, 1996, (Vol. 2)*, Mainz, September 22–25, 1996, pp. 795–799.

36. Q.-L. Li, X.-J. Ji, T. Han, and W.-K. Shi, "Walsh threshold matched-filtering based anti-collision for surface acoustic wave tags," *Journal of Shanghai Jiaotong University (Science)*, vol. 14, pp. 681–685, 2009.

37. C. Hartmann, P. Hartmann, P. Brown, J. Bellamy, L. Claiborne, and W. Bonner, "*Anti-Collision Methods for Global SAW RFID Tag Systems*," in *2004 IEEE Ultrasonics Symposium*, Montreal, August 23–27, 1–3, 2004, pp. 805–808.

38. Q. J. Teoh and N. C. Karmakar, "Anti-collision of RFID tags using capture effect," in *Handbook of Smart Antennas for RFID Systems*, Wiley Microwave and Optical Engineering Series, John Wiley & Sons, Inc., Hoboken, NJ, 2010.

39. R. E. Azim and N. Karmakar, "*A Collision Avoidance Methodology for Chipless RFID Tags*," in *Asia-Pacific Microwave Conference Proceedings (APMC), 2011*, Melbourne, VIC, December 5–8, 2011, pp. 1514–1517.

40. R. V. Koswatta and N. C. Karmakar, "*A Novel Method of Reading Multi-Resonator Based Chipless RFID Tags Using an UWB Chirp Signal*," in *Asia-Pacific Microwave Conference Proceedings (APMC), 2011*, Melbourne, VIC, December 5–8, 2011, pp. 1506–1509.

41. S. Mukherjee, "*Chipless Radio Frequency Identification by Remote Measurement of Complex Impedance*," in *European Microwave Conference, 2007*, Munich, October 9–12, 2007, pp. 1007–1010.

42. S. Preradovic, "*Chipless RFID for Barcode Replacement*," Ph.D. Thesis, Electrical and Computer Science Engineering, Monash University, Melbourne (2009).

43. I. Balbin and N. Karmakar, "*Novel Chipless RFID Tag for Conveyor Belt Tracking Using Multi-Resonant Dipole Antenna*," in *European Microwave Conference, 2009. EuMC 2009*, Rome, September 29–October 1, 2009, pp. 1109–1112.

44. I. Balbin and D. N. Karmakar, "*Radio Frequency Transponder System*," Australian Provisional Patent, DCC, Ref:30684143/DBW, Oct. 20, 2008.

45. B. Sklar and F. J. Harris, "The ABCs of linear block codes," *IEEE Signal Processing Magazine*, vol. 21, pp. 14–35, 2004.

CHAPTER 7

MULTI-TAG IDENTIFICATION THROUGH TIME–FREQUENCY ANALYSIS

7.1 INTRODUCTION

In the preceding chapter, a detailed review of anticollision and signal integrity has been presented. In this chapter a specific method of multi-tag identification based on *time–frequency* (*t–f*) analysis is presented. The popularity of chipless RFID systems for identification and tracking is increasing nowadays. It is introducing the demand of collision avoidance and multiple access in chipless RFID systems. From the reviews of the operating principle, data encoding techniques and collision in chipless RFID systems in the preceding chapter, it is evident that the available anticollision algorithms are not directly applicable for chipless RFID systems. Therefore significant research interests are developing for collision avoidance and multiple access in chipless RFID systems.

Chipless RFID systems possess more similarity with RADAR systems than the conventional RFID systems. Like RADAR targets, chipless tags backscatter the incident signal to different directions. The reader antenna captures a portion of the backscattered signal. The reader analyzes the spectrum of the signal to find the resonance information which denotes the tag ID. Figure 7.1 shows a multi-tag scenario in a

Chipless Radio Frequency Identification Reader Signal Processing, First Edition.
Nemai Chandra Karmakar, Prasanna Kalansuriya, Rubayet E. Azim and Randika Koswatta.
© 2016 John Wiley & Sons, Inc. Published 2016 by John Wiley & Sons, Inc.

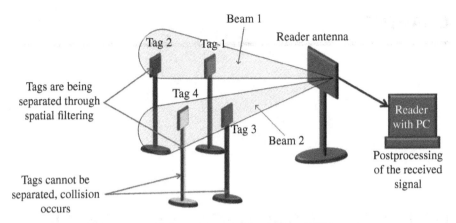

Figure 7.1 Multiple tag scenarios in chipless RFID systems.

chipless RFID system. Here, four tags are in the interrogation zone of the reader denoted as Tags 1, 2, 3, and 4. Use of a narrow-beam antenna with beam steering capability allows reading Tags 1 and 3 separately. But Tags 1 and 2 falls within the same beam (beam 1) and cause a collision. Therefore, spatial filtering with narrow-beam antenna alone is not capable of fully avoiding collision in a chipless RFID system. Therefore, the technical challenges for multiple tag detection and collision avoidance need to be addressed from signal processing perspective together with the smart antenna for employing space-division multiple access. Thus the difference in geometrical distribution of tags can be exploited as a distinguishing criterion for multiple tags. Certainly this situation will be application specific. This chapter describes a signal separation-based approach through *t–f* analysis. This method aims to extract individual tag response signals from a collided response signal.

Figure 7.2 shows the organization of the chapter. Section 7.2 introduces the *t–f* analysis, Section 7.3 describes the background theory involved in fractional Fourier transform (FrFT) and its application for multiple responses separation, Section 7.4 presents the overall system description used for validation of the method, and finally, Section 7.5 describes the simulation results for validation of the method.

7.2 *t–f* ANALYSIS AND CHIPLESS RFID SYSTEMS

As discussed in the previous chapters, frequency-domain chipless tags encode data by resonating at predefined frequencies. The identification is done by analyzing the spectrum of the response signal from the tag

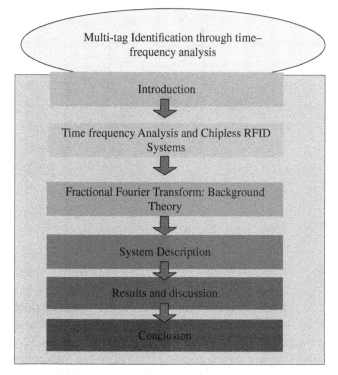

Figure 7.2   Organization of Chapter 7.

[1–5]. When multiple tags in the interrogation zone respond simultaneously to the reader, their response signals overlap each other. The collision cannot be determined directly from the spectrum of the collided signal. This may lead to a false identification.

The responses from multiple tags overlap in both time and frequency domains. Therefore neither time-domain windowing nor frequency-domain filtering is capable of separating them. Moreover, they cannot be turned off and on according to the reader's query signal to avoid collision. However, because of the spatial positions, there may be time difference of arrival (TDOA) among the response signals from multiple tags. Therefore, resonances may occur at different time instances but at the same frequency. Therefore, the time instances of tag responses (resonances) are an important parameter which may be used as a distinguishing characteristic for separating multiple tag responses. Figure 7.3 shows the *t–f* representation of response signals from Tags 1 and 2 shown in Figure 7.1. In *t–f* plane, it is possible to distinguish between the responses from two different tags. As shown in Figure 7.3, frequency f_1

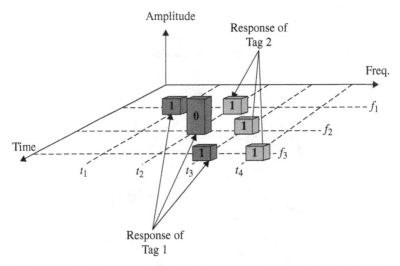

Figure 7.3   Time–frequency (t–f) plot for responses of Tags 1 and 2 in t–f plane.

is received at time instances t_1 and t_2 for Tags 1 and 2 respectively. Thus t–f analysis enables to separate them based on the times of arrival.

Different types of t–f analysis methods are used in RADAR and sonar for time-varying signal analysis [6]. There are different t–f analysis methods such as short-time Fourier transform (STFT), Wigner–Ville distribution (WVD), wavelet transform (WT), and FrFT. Here we have investigated the behavior of the response signals through STFT and FrFT. It has been found that FrFT suffers less from cross terms than STFT and also provides compact support for linearly frequency modulated (LFM) signal. Therefore, for multiple signal separation, we have used FrFT in this investigation.

The behavior of the interrogation and response signals of chipless tags in t–f domain has been investigated first. The signals have been analyzed in MATLAB through STFT. It analyzes the spectrum of the signal from the tag over the frequency as well as time. It divides the overall signal into small time windows and computes the FFT at each time window. The squared magnitude of the STFT is known as *Spectrogram* and it provides a signal's energy distribution over joint t–f domain. The difference between conventional Fourier transform (FT) and STFT is that FT calculates the frequency spectrum of the whole time-domain signal at once, hence the time information is lost [7], whereas STFT preserves the time information together with the frequency information.

For validation of the proposed method, a chipless RFID system with an interrogator and multiple tags is modeled in MATLAB. The tag is interrogated with a frequency swept signal (similar to an original interrogation system). The frequency of the interrogation swept signal is varied linearly with time creating a straight line in the t–f plane. Figure 7.4 shows the normalized power spectrum of the interrogation signal in t–f plane. Figures 7.6 and 7.8 show the normalized power of the interrogation signal over frequency and time respectively. It is evident from the figures that the frequency of the interrogation signal is varying linearly with time. Moreover, the power of the signal is constant over both time and frequency. Figure 7.5 shows the normalized power of the interrogation and response signals, respectively, in t–f domain, which is also a linearly frequency-varying signal. Figures 7.7 and 7.9 depict the response from a 4-bit chipless tag over time and frequency, respectively. They show variation in the normalized power level because of the resonances form the tag. The interrogation and response signals both are changing their frequencies linearly with time. So both the signals can be termed as LFM or chirp signals.

In a multi-tag environment, multiple tags remain in close proximity to each other; their reply signals overlap in time domain. As chipless tags operate within the same frequency bandwidth so their spectral content also overlaps. FrFT translates the signal to an intermediate domain between time and frequency where the overlapping response signals will have minimum or no overlap. From the open literature, it has been found that for LFM/chirp signals, FrFT produces the most compact support [8, 9]. As can be seen from Figures 7.6, 7.7, and 7.8, LFM signals spread over the time and frequency, but it can be represented as an impulse/spike in optimum fractional domain. This compressed representation is termed as compact here. It is more elaborately explained in the later sections. This compact property of FrFT has been exploited here for analyzing the responses of multiple chipless RFID tags and separating multiple response signals. The detailed algorithm for FrFT is presented in the next section.

7.3 FrFT: BACKGROUND THEORY

7.3.1 Linear Frequency Modulated Signal

Before describing the detail algorithm and basis of FrFT, the mathematical representation of the linear frequency modulated (LFM) signal or chirp signal is required. The LFM signal is defined by the following Equation 7.1:

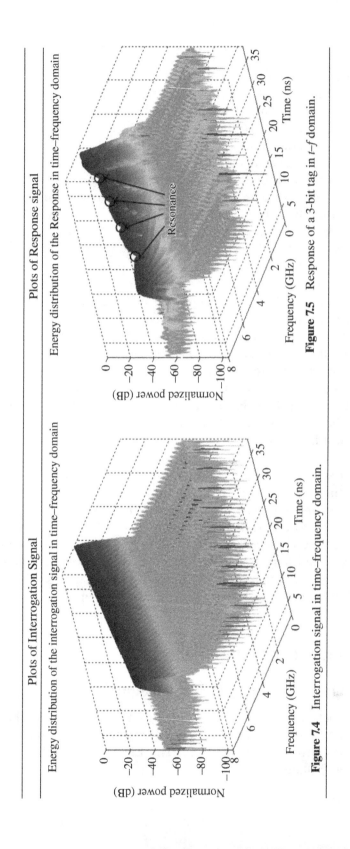

Figure 7.4 Interrogation signal in time–frequency domain.

Figure 7.5 Response of a 3-bit tag in *t–f* domain.

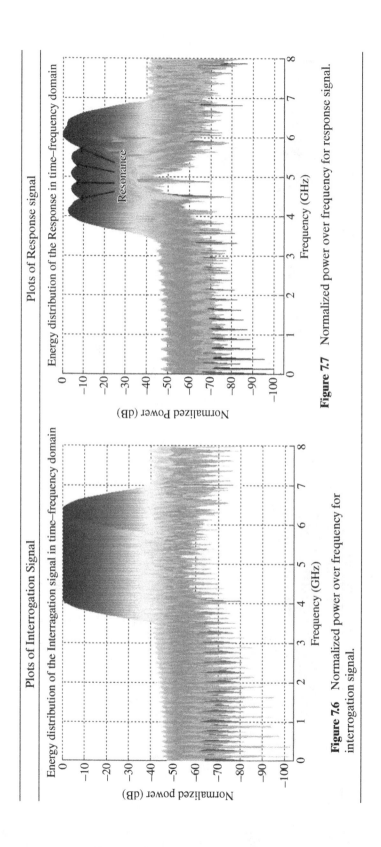

Figure 7.6 Normalized power over frequency for interrogation signal.

Figure 7.7 Normalized power over frequency for response signal.

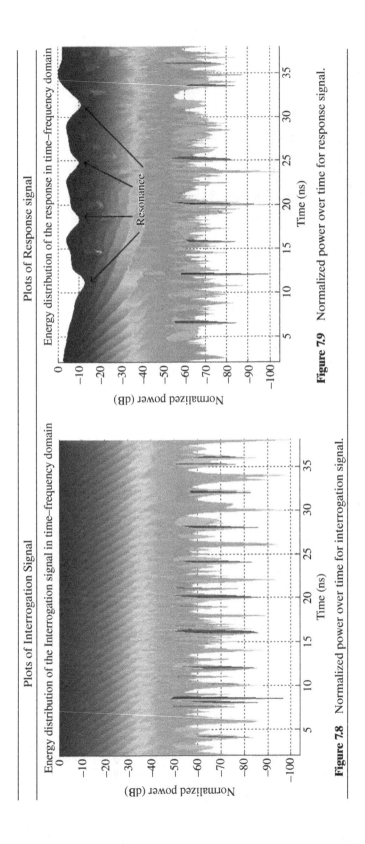

Figure 7.8 Normalized power over time for interrogation signal.

Figure 7.9 Normalized power over time for response signal.

$$x(t) = \begin{cases} e^{j2\pi\left[B/(2T)t^2 + (f_c - B/2)t\right]}, & 0 \le t \le T \\ 0 & , \text{ otherwise} \end{cases} \tag{7.1}$$

where B = total bandwidth of the chirp, T = duration of this chirp, and f_c = center frequency of sweep. The rate of change of frequency or the chirp rate is defined by

$$a = \frac{B}{T} \tag{7.2}$$

The instantaneous frequency of the signal can be calculated using the following expression:

$$f_i(t) = \frac{1}{2\pi}\varphi'(t) \tag{7.3}$$

where $\varphi(t)$ is instantaneous phase of the signal. Though expression (7.3) is used for instantaneous frequency calculation of narrowband signal, it is applicable for LFM signal as the instantaneous phase of LFM signal is differentiable. It can be interpreted as the dominant frequency at any instant. Equation 7.3 can be rewritten after inserting the value from expression (7.1) as

$$f_i(t) = \frac{d}{dt}\left(\frac{B}{2T}t^2 + \left(f_c - \frac{B}{2}\right)t\right) = \frac{B}{T}t + f_c - \frac{B}{2} \tag{7.4}$$

Expression (7.4) confirms the linear relationship between time and frequency. It is graphically represented in Figure 7.10:

7.3.2 FrFT

FrFT is an extension of FT. It can be termed as a generalization of the FT [8, 10]. It can be considered as an angle rotation in the t–f plane where the signal contains both time and frequency information unlike the time-domain representation or spectrum of a signal. The definition of $f(t)$ in FrFT domain is given by Equation 7.5:

$$f_p(u) = \int_{-\infty}^{\infty} f(t)K(\alpha; u, t)dt \tag{7.5}$$

where p = order of FrFT and α = rotational angle between time axis and FrFT axis. FrFT is actually a rotation of the t–f plane of a signal, which creates a t–f mapping of a signal whose frequency is varying with time. Being a linear transform, it suffers less from cross terms than other t–f analysis method [11]. Figure 7.11 shows the rotation of t–f plane through FrFT.

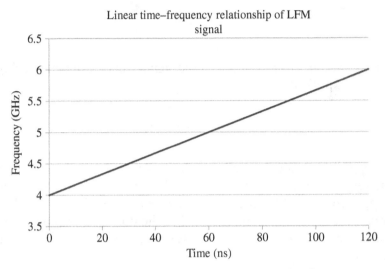

Figure 7.10   Time-frequency plot of LFM signal.

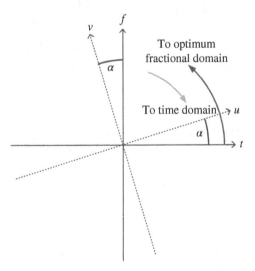

Figure 7.11   Rotation of t–f plane through FrFT.

The relationship between rotational angle and order of transformation is shown by Equation 7.6.

$$p = \frac{2}{\pi}\alpha \tag{7.6}$$

The kernel $K(\alpha;u,t)$ is defined as in Equation 7.7:

$$K(\alpha;u,t) = \sqrt{\frac{1 - i\cot\alpha}{2\pi}}\exp\left[i\left(\frac{u^2 + t^2}{2}\cot\alpha - ut\csc\alpha\right)\right] \tag{7.7}$$

The transformation can be described through four steps as shown in Figure 7.12 [12–14]:

- *Multiplication by a chirp in one domain*
- *Fourier transformation*
- *Multiplication by chirp in the transformed domain (complex exponentials with linear frequency modulation)*
- *Complex scaling*

Certain transformation orders are particularly notable as they are being identical to mathematical transforms [15] (Table 7.1).

7.3.2.1 LFM Signal in Fractional Domain

FrFT is considered as a rotation by an angle in the *t–f* plane or a decomposition of a signal in terms of chirps [7]. A chirp forms a line in the *t–f* plane; therefore, there exists an order of transformation where such signal is compact as chirp is neither compact in time domain nor in frequency domain.

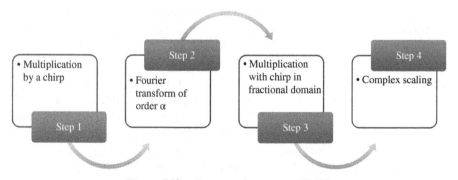

Figure 7.12   Computation steps of FrFT.

Table 7.1 Transformation Order and Relation to Transform

Order of Transformation	Rotational Angle	
$p=0$	$\alpha=0$	Time-domain signal
$p=1$	$\alpha=\pi/2$	Fourier transform (FT)
$p=-1$	$\alpha=-\pi/2$	Inverse Fourier transform
$0<p<1$	$0<\alpha<\pi/2$	Fractional Fourier transform

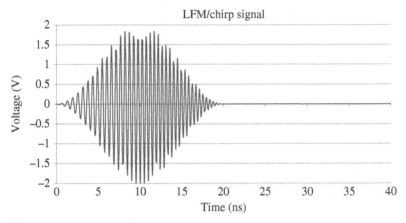

Figure 7.13 LFM/chirp signal (time domain).

This is the reason that FrFT produces the most compact support for linear chirp signal (LFM Signal) in the optimum fractional domain [8].

Figure 7.13 shows an LFM signal in the time domain whose frequency varies from 1.5 to 2.5 GHz in 20 ns. The spectrum is shown in Figure 7.14. The Hanning window has been used on the time-domain signal to suppress the side lobes. The Hanning window is also known as *raised cosine* window. Next, the time-domain signal has been transformed to the optimum fractional domain using FrFT transformation and the result is shown in Figure 7.15. It is evident from Figure 7.15 that the LFM signal creates an impulse like shape in optimum fractional domain compared to time- or frequency-domain representation. It is termed as "*compact.*" This property is being exploited for separating of multiple overlapping LFM signals in various literatures [12, 16–18]. The detail transformation procedure and calculation of optimum transformation order for a particular chirp signal will be discussed in Section 7.3.2.2. It is evident from the figure.

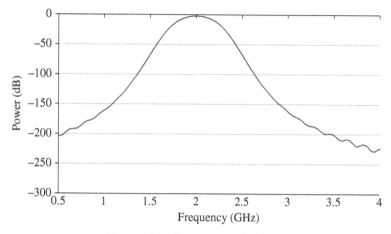

Figure 7.14 Spectrum of the chirp.

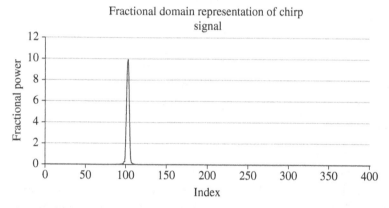

Figure 7.15 Compact representation of LFM signal in fractional domain.

7.3.2.2 Implementation of Discrete FrFT for Multiple Overlapping Signal Separation

In practice, the discrete FrFT is calculated. Here the implementation algorithm described in Refs. [19, 20] has been used. The optimum order of transformation of an LFM signal depends on the rate of change of frequency of the LFM signal df/dt. It actually defines how much rotation of the t–f plane is required for the compact representation of the LFM signal in the fractional domain. The optimum transform order is defined by Equation 7.8 [15]:

$$p_{\mathrm{opt}} = -\frac{2}{p}\tan^{-1}\left(\frac{1}{2a}\right) \tag{7.8}$$

where chirp rate $a = B/T$, B = bandwidth of the LFM signal, and T = total duration of the signal. The input signal is discrete and sampled according to Nyquist sampling theory. To calculate the order p_{opt} requires the knowledge of time and frequency resolution dt and df. Hence Equation 7.8 is further modified as

$$P_{opt(discrete)} = -\frac{2}{p}\tan^{-1}\left(\frac{df/dt}{2a}\right) \tag{7.9}$$

In expression (7.9) $df = f_s/N$ and $dt = 1/f_s$, N = total number of samples in the waveform. This equation has been implemented and the algorithm described in Refs. [19,20] for implementing FrFT for multiple response signal separation in chipless RFID system.

To validate the idea, a simulation has been carried out in MATLAB where an analytic chirp signal has been generated and then delayed. Then the original and delayed version has been added together with significant overlap between them. Then the combined signal was transformed to fractional domain. The results are being presented here. The simulation parameters used are shown in Table 7.2.

Figure 7.16 shows two overlapping chirp signals with same center frequency bandwidth but delayed by a small amount about $\tau = 8$ ns from each other. The frequency spectrums of the signals are shown in Figure 7.17. The t–f representation of the signals is shown in Figure 7.18. It is evident from the figure that signals are overlapping in both time and frequency domain. FrFT rotates the t–f plane to an optimum angle where the signals are having minimum or no overlap. The transformation order is calculated from expression (7.9). Figure 7.19 shows the overlapped chirp/LFM signals in optimum fractional domain. The two signals are concentrated at two different regions in the optimum domain. Now windowing in this domain enables to separate them from each other.

As already discussed the response signal from chipless tag is also an LFM/chirp signal with variation in power level due to resonances from

Table 7.2 Value of Variables Used in Simulation

Sampling frequency, f_s	10 GHz
No. of data points, N	401
Bandwidth, B	2 GHz
Chirp duration, T	20 ns
Order of transformation, p_{opt}	−0.8737
Corresponding angle of rotation	−78.633°

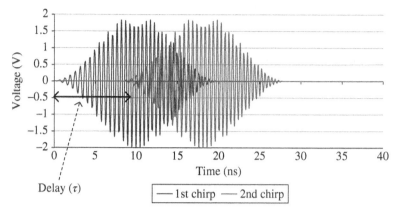

Figure 7.16 Two overlapped chirp signals.

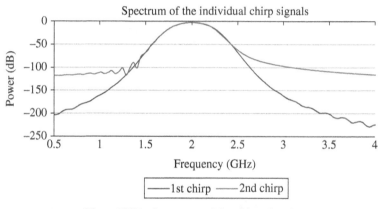

Figure 7.17 Spectrum of the chirp signal.

tags. Therefore, for collided response signal, the scenario is comparable to the one described previously. The previous analysis creates a foundation for using FrFT as a signal separation method for multiple responses from chipless tags.

7.4 SYSTEM DESCRIPTION

Here, for analyzing the TDOA-based signal separation approach through FrFT, we are considering a simulation setup where multiple tags are in the interrogation zone but at different geometrical positions. Therefore, their response signals are having delays from one another.

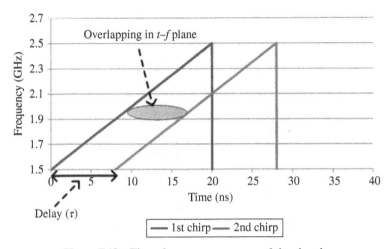

Figure 7.18   Time–frequency response of the signals.

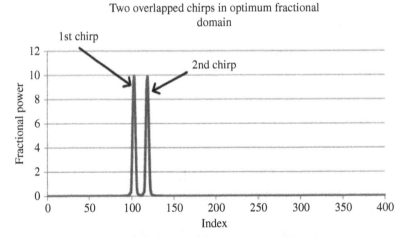

Figure 7.19   The overlapped chirps in optimum fractional domain.

Two examples of such scenarios are shown in Figure 7.20. In Figure 7.20a, Tags 1 and 2 are having distances r_1 and r_2, respectively, from the receiving antenna. In Figure 7.20b, two tags are on a conveyer belt and responding simultaneously to the reader. They are having a physical distance of R between them. Therefore, in both cases, we are considering a scenario where spatial filtering through beam steering with a narrow-beam antenna cannot separate the response signals. These situations demand a signal separation-based approach to solve the collision problem.

As discussed already, the LFM/chirp signal has excellent t–f relationship. Therefore we have used an LFM signal as the interrogation signal to utilize the t–f relationship of the signal for signal separation. Figure 7.21 shows the representation of the interrogation and response signals in t–f plane when interrogated with an LFM signal. The signals create straight lines in the t–f plane. As the tags are in close proximity to each other, before Tag 1 completes sending the modulated signal, another tag starts responding. Therefore, a collided signal is received by the reader antenna. As can be seen in Figure 7.21, because of the positional variation between the tags, there is a delay τ between the response signals. FrFT utilizes this initial delay between them to

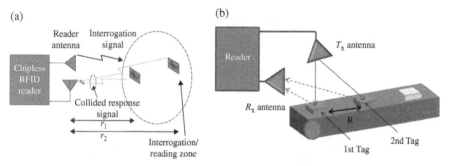

Figure 7.20 Chipless RFID system, multiple tags in the interrogation zone. (a) Multiple tag reading in a room and (b) multiple tags in conveyer belt.

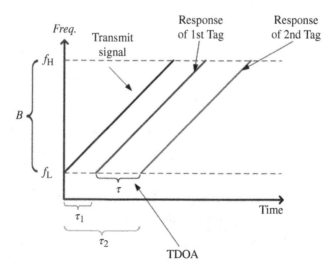

Figure 7.21 Overlapping response signals from chipless tags in t–f plane.

transform the signals to the optimum domain where they have minimum or no overlap. Now, the optimum angle of rotation, α, is calculated from expression (7.6). As seen in Figure 7.22, the rotation of the t–f plane to an angle α results in a nonoverlapping situation among the signals. Now they can be windowed separately to individually separate each signal.

The simulation to verify the postprocessing method has been carried out in both ADS 2009 and MATLAB 2010 environment. The chipless RFID reader and tags have been modeled in ADS 2009 and postprocessing of the data has been performed in MATLAB 2010. The block diagram of the total simulation is presented in Figure 7.23.

7.4.1   ADS Simulation Environment

In ADS 2009, a series of band-stop filters have been used in series to model the chipless RFID tag [21]. Here, two different sets of simulation have been performed. In one, two 1-bit tags have been simulated where both the tags resonate around 3.95 GHz. Hence, the band-stop filter was set to resonate at 3.95 GHz. In another simulation 3-bit tags have been modeled with three band-stop filters resonating at three different frequencies. Two antennas have been used with the band-stop filters, one for receiving the interrogation signal from the reader antenna and the other one for retransmitting the signal after modulation from the tag. The antennas are assumed to have unity gain for simplicity. Here, the presence and absence of a resonator (attenuation of power at the resonance frequency) is decoded as logic "1" and "0," respectively. An inbuilt source is used for generating the interrogation chirp signal.

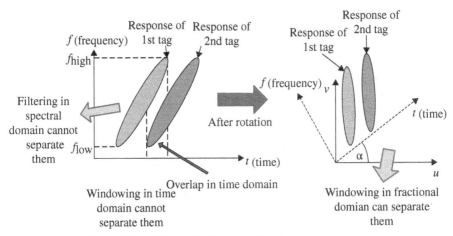

Figure 7.22   Effect of FrFT on collided response signals.

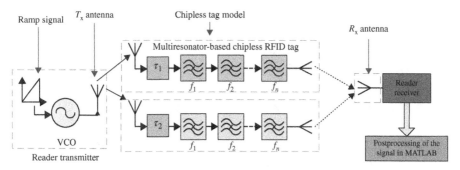

Figure 7.23 Block diagram of ADS 2009 simulation setup and postprocessing in MATLAB.

To create a spatial separation and hence a delay between the response signals of two tags, a delay unit is used with each tag. A power splitter splits the incoming power into two equal sections to feed to two tags. The reader transmitting part contains an LFM generator and a unity gain antenna. The receiving part also contains an antenna, an LNA for amplifying the received signal.

7.4.2 Postprocessing in MATLAB

The postprocessing on the received signal that includes separation of the collided signals and extraction of tag IDs is performed in MATLAB through signal processing. The flowchart of the postprocessing steps is presented in Figure 7.24.

The interrogation signal is $x(t)$, which is an LFM signal expressed by expression (7.1). We are considering two tags in the interrogation zone. The response signals from the tags are expressed as

$$y_1(t) = H_{\text{tag1}}\left[x(t - \tau_1)\right] \tag{7.10}$$

$$y_2(t) = H_{\text{tag2}}\left[x(t - \tau_2)\right] \tag{7.11}$$

where H_{tag} is the transfer function of the chipless tag, which makes amplitude and phase variation in the response signal. $y_1(t)$ and $y_2(t)$ are the response signals from Tags 1 and 2, respectively, which are received by the reader after τ_1 and τ_2 initial delay compared to interrogation signal where $(\tau_1 - \tau_2) = \tau$. As the tags are in close proximity to each other, $y_1(t)$ and $y_2(t)$ overlap in time domain, and the received signal by the receiving antenna is expressed as

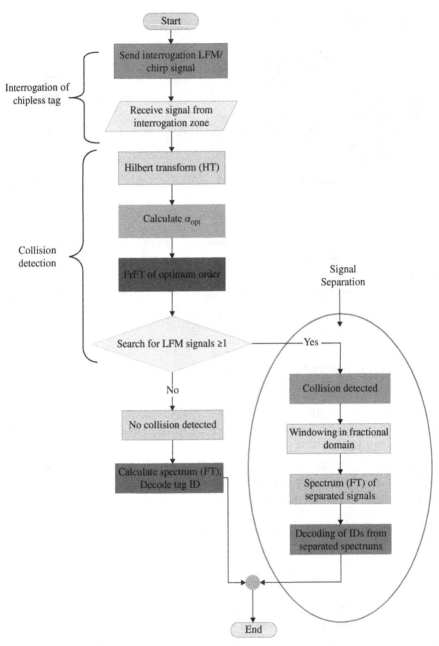

Figure 7.24 Postprocessing flowchart (MATLAB).

$$y(t) = y_1(t) + y_2(t) = H_{\text{tag1}}\left[x(t-\tau_1)\right] + H_{\text{tag2}}\left[x(t-\tau_2)\right] \quad (7.12)$$

For simplicity, we have considered a noise-free propagation of signals. As the retransmission-based tag [22] is considered here, the interrogation and response signals will be in opposite polarity. This will separate the response signals from the interrogation signal. The received signal $y(t)$ is converted to an analytic form through Hilbert transform. The next step is to calculate the optimum angle of rotation α_{opt} for converting $y(t)$ to the optimum fractional domain. However, as from expressions (7.6) and (7.8), it is evident that the required parameter is the chirp rate of the received signal. However, as the received signal is a delayed and modulated version of the interrogation one, the chirp rate remains same. Next, $y(t)$ is converted to optimum fraction domain and expressed as

$$Y_{p_{\text{opt}}}(u) = \int_{-\infty}^{\infty} y(t) K(\alpha; u, t) dt \quad (7.13)$$

Next, the regions for signals $y_1(t)$ and $y_2(t)$ are searched in the fractional domain. Then individual signals are windowed out from fractional domain using a rectangular window of appropriate size:

$$w(u) = \begin{cases} 1, & u_1 \leq u \leq u_2 \\ 0, & \text{elsewhere} \end{cases} \quad (7.14)$$

where $u_1 \leq u \leq u_2$ is the region of individual LFM signals in the fractional domain. The individual response signals are extracted using the following expressions:

$$Y_1(u) = w_1(u) \times Y_{p_{\text{opt}}}(u) \quad (7.15)$$

$$Y_2(u) = w_2(u) \times Y_{p_{\text{opt}}}(u) \quad (7.16)$$

Now, for decoding individual tag IDs, the separated signals are converted back to time domain from fractional domain by rotating the individual signals t–f plane in opposite direction but the same amount, α_{opt}. The time-domain signals are restored by performing an inverse FrFT (iFrFT) or FrFT of order $-p_{\text{opt}}$ on $Y_1(u)$ and $Y_2(u)$ separately. It is expressed by the following expressions:

$$y_{1,\text{retored}} = F_{-p_{\text{opt}}}\left(Y_1(u)\right) \quad (7.17)$$

$$y_{2,\text{retored}} = F_{-p_{\text{opt}}}\left(Y_2\left(u\right)\right) \tag{7.18}$$

The tag IDs are decoded from the spectrum of $y_{1,\text{restored}}$ and $y_{2,\text{restored}}$. The spectrums of the individual signals are calculated using FT. The detailed explanation of the separation process is explained with simulation results in the next section.

7.5 RESULTS AND DISCUSSION

As stated previously, the first simulation is performed to separate two one-bit tag signals where both the tags resonate at 3.95 GHz. The composite signal is shown in Figure 7.25. The signal contains two LFM signals that have a delay of 20 ns between them. The ID of a chipless tag is decoded from the spectrum of the received signal. But the spectrum of the composite signal does not provide any useful information regarding the tag ID as there are constructive and destructive interference between the signals. This situation makes it impossible to decode the ID from the spectrum of this signal. The composite signal $y(t)$ is shown in Figure 7.25. The numeric values of the parameters that are used to carry out the simulation are outlined in Table 7.3.

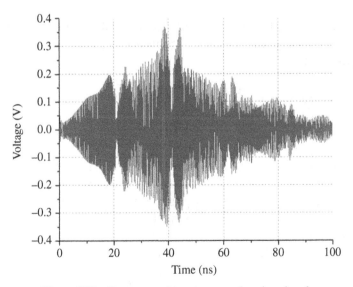

Figure 7.25 Response of two tags overlapping signal.

Table 7.3   **Specifications of the Variables**

Bandwidth, B	1 GHz
Chirp duration, T	40 ns
Chirp rate, a	25 MHz/ns
Order of transformation, p_{opt}	−0.5378
Corresponding angle of rotation, α	−48.402°
Delay between two tags, τ	20 ns

This signal is imported from ADS to MATLAB for further analysis for signals separation of individual tag responses. The step-by-step analysis procedure is described below.

Stage 1: Transformation of Time Domain Response to Fractional Domain

The rotation of the t–f plane by an optimum angle forms a fractional plane where the individual LFM components become perpendicular to the fractional axis. The rate of change of frequency or the chirp rate is calculated from expression (7.2) as the bandwidth B and the chirp duration T are known beforehand. The chirp rate defines the value of rotation of the t–f plane according to expression (7.8). As the data is in discrete form, that means time sampled, we use the discrete calculation of the order of transformation, which is calculated by expression (7.9). For this particular simulation, the order of transformation is −0.5378. Applying FrFT maximally compresses the overlapping signals in fractional domain with peaks and nulls. The transformed signal $Y_{p_{opt}}$ is shown in Figure 7.26.

Stage 2: Estimation of Number of Tags and Windowing in Fractional Domain

As already discussed, windowing in fractional domain is the best solution to separate the individual response signals as seen from Figure 7.26. For estimating the number of tags and windowing, first the peaks need to be detected and then the minimum points on either sides of the peak are detected. The additional peaks and nulls occur as the individual chirps contain the frequency signature of an individual tag within the spectrum. This creates a challenging situation in detecting the number of peaks. But the minimum points on either side of a chirp give the indication about the starting and ending points of a component as those minima show less value than the nulls. Therefore,

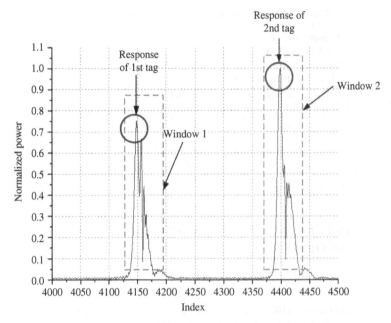

Figure 7.26 Collided signals in fractional domain.

rather than only detecting the peaks, the minimum nulls on either side of the peak need to be detected. The windowing consists of finding the minimum points and then extracting only this portion while making other portions zero. A rectangular window expressed in Equation 7.14 is placed around each component and then they are separated from the fractional domain. The individual separated signals are denoted by $Y_1(u)$ and $Y_2(u)$.

Stage 3: Decoding the Individual Tag ID

Windowing in fractional domain separates each tag response. The tag ID needs to be decoded from the spectral component. It can be done in two ways:

- The individual signals can be restored in time domain by an Inverse FrFT of the order $-p_{\mathrm{opt}}$, which will restore the signals in the time domain.
- Another way is to directly convert the signal from optimum domain to frequency domain using the same FrFT algorithm where the order of transformation would be $(1 - p_{\mathrm{opt}})$.

In either way, the spectrum of the individual signal is calculated and from the spectrum dip at predefined frequency range is searched, which is the ID of the tag. Figure 7.27 shows the spectrum. The black curve is the spectrum of the collided signal, which does not provide any useful information. The gray curve is the spectrum of the 1st tag and the black is the spectrum of the 2nd tag. Both shows resonance around 3.95 GHz as expected. Similar algorithm has been applied for few other combinations of tag for less delay between the signals.

Figures 7.28 and 7.30 show the spectra of two tags for two different combinations of tag IDs—"000," "101" and "111," "000." In each case, the delay between response signals of Tag 1 and Tag 2 is 3 ns, which corresponds to a distance of 0.9 m. The signals are separated in fractional domain and then converted to the spectral domain, and finally, the tag ID is decoded from the spectrum.

Figures 7.29 and 7.30 show the spectral signatures and decoded IDs for two tags with IDs "111" and "010." The spectrum shows three nulls corresponding to three resonances for Tag 1. For Tag 2 there should be only one resonance. However, the spectrum shows a dip at

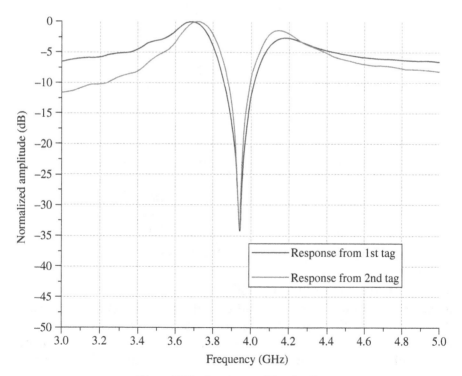

Figure 7.27 Spectrum of the signals.

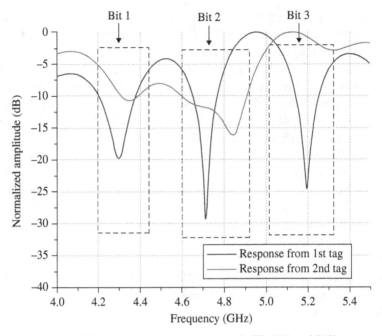

Figure 7.28 Spectrum of two tag signals (ID: 111 and 010).

Freq. (GHz)	Resonance		Bit	
	Tag 1	Tag 2	Tag 1	Tag 2
4.2–4.42	Yes	No	1	0
4.6–4.9	Yes	Yes	1	1
5–5.3	Yes	No	1	0

Decoded correctly

Figure 7.29 Table of result, IDs correctly decoded.

about 4.3 GHz and another one at around 4.85 GHz. As the first one is not as prominent as the second one, the first one is not an actual dip. Because of the tag collision, this spurious dip is there. As the delay between the signals is very small, only 3 ns, the rotation of the t–f plane is not always fully capable of removing the interference from other tag's response signal.

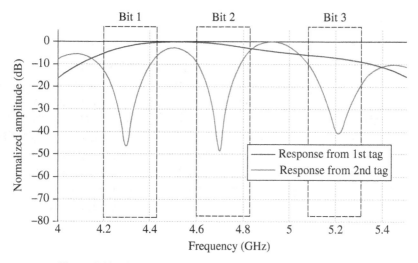

Figure 7.30   Spectrum of two tag signals (ID: 000 and 111).

Freq. (GHz)	Resonance		Bit	
	Tag 1	Tag 2	Tag 1	Tag 2
4.2–4.42	Yes	No	1	0
4.6–4.9	Yes	No	1	0
5–5.3	Yes	No	1	0

Decoded correctly

Figure 7.31   Table of result, IDs correctly decoded.

Figures 7.30 and 7.31 show the spectral signatures and decoded tag IDs for two tags with IDs "111" and "000." The spectrum of Tag 1 shows three clear dips, and the spectrum of Tag 2 does not have any dip. So the response signals are successfully separated.

From the aforementioned comprehensive investigation for two multi-bit multi-tag scenarios, it can be inferred that the proposed method is capable of identifying individual tags in multiple tag reading.

7.6 CONCLUSION

In this chapter, a novel signal separation methodology has been presented for multiple chipless tags identification. This method uses FrFT for analyzing and separating multiple tag responses in t–f plane. The background theory of FrFT has been presented together with the potentiality of this method for multiple tag reading. The method has been investigated for different combinations of multi-bit chipless RFID tags. In each case, this method has successfully extracted individual tag response signals from the collided signal. From the extracted signal, the tag ID is decoded correctly in each case. Therefore, this t–f analysis method can be successfully used for multiple tag identification for chipless RFID tags.

REFERENCES

1. S. Preradovic and N. C. Karmakar, "Chipless RFID: bar code of the future," *IEEE Microwave Magazine*, vol. 11, pp. 87–97, 2010.
2. S. Preradovic and N. C. Karmakar, "*Multiresonator based chipless RFID tag and dedicated RFID reader*," in *IEEE MTT-S International Microwave Symposium Digest (MTT), 2010*, Anaheim, CA, May 23–28, 2010, pp. 1520–1523.
3. I. Balbin, "*Multi-bit Fully Printable Chipless Radio Frequency Identification Transponders*," Ph.D., Electrical and Computer Science Engineering, Monash University, Melbourne.
4. I. Balbin and D. N. Karmakar, "*Radio Frequency Transponder System*," Australian Provisional Patent, DCC, Ref:30684143/DBW, October 20, 2008.
5. I. Balbin and N. Karmakar, "*Novel chipless RFID tag for conveyor belt tracking using multi-resonant dipole antenna*," in European *Microwave Conference, 2009. EuMC 2009*, Rome, September 29–October 1, 2009, pp. 1109–1112.
6. B. Boashash, Ed., *Time Frequency Signal Analysis and Processing: A Comprehensive Reference*. Amsterdam, the Netherlands/Boston: Elsevier, 2003.
7. A. S. Amein and J. J. Soraghan, "Azimuth fractional transformation of the fractional chirp scaling algorithm (FrCSA)," *IEEE Transactions on Geoscience and Remote Sensing*, vol. 44, pp. 2871–2879, 2006.
8. C. Capus and K. Brown, "Fractional Fourier Transform of the Gaussian and fractional domain signal support," *IEEE Vision, Image and Signal Processing*, vol. 150, pp. 99–106, 2003.

9. C. Wang, Z. Zhang, and S. Li, "*Interference avoidance using Fractional Fourier Transform in transform domain communication system*," in *The 9th International Conference on Advanced Communication Technology*, Phoenix, February 12–14, 2007, pp. 1756–1760.

10. A. C. McBride and F. H. Kerr, "On Namias's fractional Fourier transforms," *IMA Journal of Applied Mathematics*, vol. 39, pp. 159–175, 1987.

11. J. Bai, M. Gao, and C. Xu, "*Method of stepped frequency signal processing based on Fractional Fourier Transform*," in *9th International Conference on Signal Processing, 2008. ICSP 2008*, Beijing, October 26–29, 2008, pp. 2550–2553.

12. C. Capus, Y. Rzhanov, and L. Linnett, "*The analysis of multiple linear chirp signals*," in *IEE Seminar on Time-scale and Time-Frequency Analysis and Applications (Ref. No. 2000/019)*, London, February 29, 2000, pp. 4/1–4/7.

13. L. B. Almeida, "The fractional Fourier transform and time-frequency representations," *IEEE Transactions on Signal Processing*, vol. 42, pp. 3084–3091, 1994.

14. S. Jang, W. Choi, T. K. Sarkar, M. Salazar-Palma, K. Kyungjung, and C. E. Baum, "Exploiting early time response using the fractional Fourier transform for analyzing transient radar returns," *IEEE Transactions on Antennas and Propagation*, vol. 52, pp. 3109–3121, 2004.

15. C. Capus, "Short-time fractional Fourier methods for the time-frequency representation of chirp signals," *The Journal of the Acoustical Society of America*, vol. 113, p. 3253, 2003.

16. D. M. J. Cowell and S. Freear, "Separation of overlapping linear frequency modulated (LFM) signals using the fractional Fourier transform," *IEEE Transactions on Ultrasonics, Ferroelectrics and Frequency Control*, vol. 57, pp. 2324–2333, 2010.

17. S. Xue-jun, W. Rong-hui, and Q. Xin, "*A new multiple-access method based on Fractional Fourier Transform*," in *Canadian Conference on Electrical and Computer Engineering, 2009. CCECE '09*, St. John's, NL, May 3–6, 2009, pp. 856–859.

18. S. A. Elgamel and J. Soraghan, "*A new Fractional Fourier Transform based monopulse tracking radar processor*," in *2010 IEEE International Conference on Acoustics Speech and Signal Processing (ICASSP)*, Dallas, TX, March 14–19, 2010, pp. 2774–2777.

19. C. Candan, M. A. Kutay, and H. M. Ozaktas, "The discrete fractional Fourier transform," *IEEE Transactions on Signal Processing*, vol. 48, pp. 1329–1337, 2000.

20. H. M. Ozaktas, O. Arikan, M. A. Kutay, and G. Bozdagt, "Digital computation of the fractional Fourier transform," *IEEE Transactions on Signal Processing*, vol. 44, pp. 2141–2150, 1996.

21. S. Preradovic, I. Balbin, N. C. Karmakar, and G. Swiegers, *"Chipless frequency signature based RFID transponders,"* in *European Conference on Wireless Technology, 2008. EuWiT 2008*, Amsterdam, October 27–28, 2008, pp. 302–305.

22. S. Preradovic, I. Balbin, N. C. Karmakar, and G. F. Swiegers, "Multiresonator-based chipless RFID system for low-cost item tracking," *IEEE Transactions on Microwave Theory and Techniques*, vol. 57, pp. 1411–1419, 2009.

CHAPTER 8

FMCW RADAR-BASED MULTI-TAG IDENTIFICATION

8.1 INTRODUCTION

In Chapter 6, two different multi-tag identification methodologies have been introduced for chipless RFID systems. The first one is the time–frequency analysis using fractional Fourier transform (FrFT). In Chapter 7, FrFT-based t–f analysis has been explained in detail with the theoretical background and some preliminary results. However, it has been described that this methodology requires a short-duration chirp signal and high sampling rate [1], whereas the second method—FMCW RADAR-based multi-tag identification—requires lower sampling rate and can perform satisfactorily even with longer-duration chirp signals. The added feature is that the range of each tag in the interrogation zone can also be identified [2]. Therefore, in this chapter, the FMCW RADAR-based technique has been explored in detail for multiple chipless tag identification.

Before going to a detailed analysis of the technique, first, we need to identify the similarities and differences between RADAR and chipless RFID systems. This will help to understand the potentiality of the FMCW technique to be used in chipless RFID systems. As we all know,

Chipless Radio Frequency Identification Reader Signal Processing, First Edition.
Nemai Chandra Karmakar, Prasanna Kalansuriya, Rubayet E. Azim and Randika Koswatta.
© 2016 John Wiley & Sons, Inc. Published 2016 by John Wiley & Sons, Inc.

RADAR stands for radio detection and ranging. Radar detects the presence of targets, locates their position in space, and determines the velocity and the direction of the target's movement by transmitting EM energy and observing the returned echo as shown in Figure 8.1a [2]. Chipless RFID is a microscopic form of passive RADAR, which not only explores the presence, position, and velocity but also the identification of the target [3]. For frequency signature-based chipless tag identification, a wideband EM signal is transmitted toward the interrogation zone, and the returned echo from the tag is analyzed in the frequency domain to find resonances at predefined frequencies as shown in Figure 8.1b.

Though the interrogation and echo receiving part is similar in both RADAR and chipless RFID systems, the main focus of RADAR is to find the presence of a target and locate the range. In chipless RFID systems, the identification of chipless tag is done from resonances in the echo signal. So the returned echo is analyzed more precisely for extracting special features of the tag (resonances) from it [4, 5]. The FMCW technique in RFID has been explored for reading chipless SAW tags [6, 7]. However, the potentiality of the method for multiple frequency signature-based chipless tag identification has not been

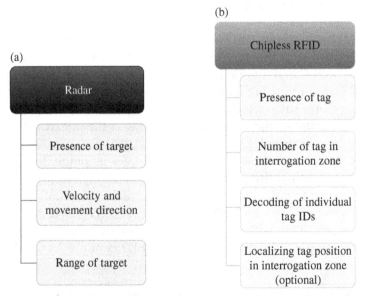

Figure 8.1 (a) RADAR's operating goals. (b) Chipless RFID system's operating goals.

addressed earlier. The chapter is all about detection of a chipless RFID tag using FMCW RADAR technique.

The organization of Chapter 8 is shown in Figure 8.2. The introduction section presents a comparison between the two proposed methods—FRFT-based t–f analysis and FMCW RADAR followed by similarities and differences of chipless RFID system and conventional RADAR. Section 8.2 describes the background theory of FMCW technique and the theoretical aspects for using this method in chipless RFID system. Section 8.3 provides a detailed description of the simulation environment and postprocessing method. Section 8.4 includes the results; Section 8.5 is about the discussion of the results followed by conclusion in Section 8.6.

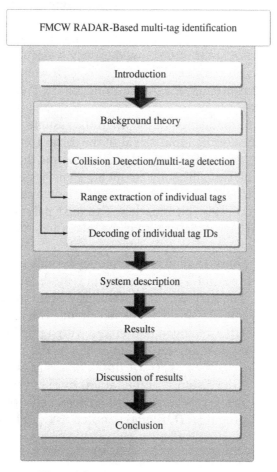

Figure 8.2 Organization of Chapter 8.

8.2 BACKGROUND THEORY

FMCW RADAR technique is used to discriminate multiple targets in a radio interrogation zone [2, 8–10]. In chipless RFID systems, this technique is used to detect and discriminate proximity tags. In this section, first the background theory of FMCW RADAR method is presented for a single target environment. Later in this section, the FMCW technique has been explained for multi-tag identification in chipless RFID systems.

8.2.1 Overview of FMCW RADAR

Figure 8.3a and b shows the block diagram of monostatic and bistatic FMCW RADAR, respectively. In a monostatic configuration, the same antenna is used for transmitting interrogation signal and receiving echo from target. A circulator is used before the antenna, which prevents the received signal to come to the transmitter section as shown in Figure 8.3a [2, 11]. A bistatic RADAR is the one that uses a transmitter and receiver separated by a considerable distance [2, 12]. However, any RADAR that uses separate antennas for transmitting and receiving can be called bistatic RADAR [2]. In this scenario, the signal received by the receiving antenna is brought to the mixer. In both of the systems, a coupler is used to bring a reference of the input signal to the mixer for comparison. The mixer crates a difference frequency signal comparing the reference signal and received signal. This difference frequency signal is called intermediate frequency (IF) signal. The IF signal is further analyzed for extracting target velocity and range information. The next paragraph explains in detail about the IF signal generation in the mixer and information extraction from the IF signal.

A constant amplitude linearly frequency-modulated signal $x(t)$ is transmitted for a period T_R. The frequency modulation is used as a timing mark here. The frequency is continuously swept from a lower value of f_1 to a higher value of f_2 for N frequency points. The reflection or echo from the target, $y(t)$, is received and mixed with a portion of the transmitted signal, $x_{ref}(t)$, forming an IF signal, $y_{IF}(t)$. The transmitter sends the signal with frequency f_1, $x_{f_1}(t)$ at time t_1 and signal with f_2, $x_{f_2}(t)$ at time t_2. Because of the distance of the target, $x_{f_1}(t)$ requires a finite time to hit the target and travel back to the receiver. This time is called as round-trip time of flight ($RTOF$) and denoted as t_{RTOF}. It is expressed as

$$t_{RTOF} = \frac{(2 \times R)}{c} \tag{8.1}$$

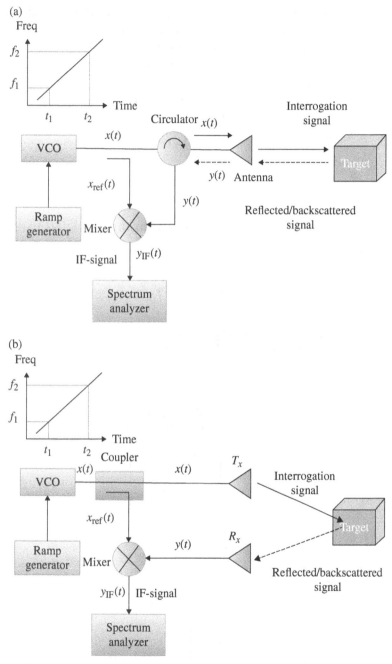

Figure 8.3   (a) Block diagram of monostatic FMCW technique. (b) Block diagram of bistatic FMCW technique.

where R is the one-way distance of the target and c is the speed of light in free space (3×10^8 m/s). Figure 8.4a and b shows the transmitted and received signals and mixer output, respectively, in the time–frequency plane. When $x_{f_1}(t)$ reaches to the receiver, the VCO generates a different frequency signal, say, as $x_{f_i}(t)$, where $i = 1, 2, ..., N$. Hence, when they are mixed in the mixer, the output is the difference frequency Δf. This low-frequency signal is IF signal. It is represented by the following expression [9]:

$$f_{IF} = \frac{t_{RTOF} \times B}{T_R} \tag{8.2}$$

The range of the target is calculated from the IF signal using the following expression:

$$R = \frac{c \times T_R}{2 \times B} \times f_{IF} \tag{8.3}$$

Here, R is the range of the target and B is the bandwidth of the transmitted signal, which is expressed as $B = f_2 - f_1$.

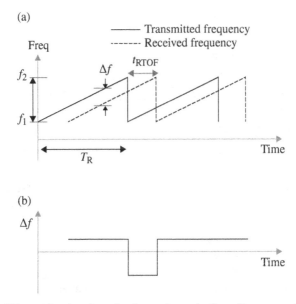

Figure 8.4   (a) Transmitted and received waveforms in time–frequency plane. (b) Mixer output in time–frequency plane.

8.2.2 FMCW RADAR Technique for Chipless RFID Systems: Multi-Tag Identification

The chipless RFID reader for multi-tag identification needs to be application specific. For different applications of chipless RFID the requirement of reader varies. For example, in some applications, a hand-held reader is desirable, whereas in some other applications, a static reader is more appropriate. In the case of handheld reader, the main requirements are lightweight and easy to carry, and in such cases, we may not have the luxury of using two different antennas for transmitting the interrogation signal and receiving response from tags. Therefore, in this scenario, single antenna can be used for transmission and reception or two antennas collocated at the same position creating monostatic radar configuration for the reader. For single-antenna configuration, a shared aperture [13] technique may be used. In a shared aperture technique, the aperture of a single array antenna can be shared through time multiplexing or polarization diversity. However, in chipless RFID systems, the small delay between interrogation and response signals makes it challenging to use the same aperture through time sharing. Therefore, polarization diversity sharing same aperture of antenna may be an acceptable solution. As shown in Figure 8.5, for library tagging

Figure 8.5 Handheld reader for library tagging system.

system, a lightweight, handheld reader is desirable where using monostatic RADAR configuration with single antenna is desirable.

However, there are applications where the bistatic RADAR configuration may be advantageous as shown in Figure 8.6. It represents a vehicle entry monitored with RFID system. In this scenario, the signal from the tag will be scattered in various direction, not only in the direction of the transmitter. Therefore, using of multiple receivers at different locations is advantageous for accurate identification.

Therefore, the use of mono- or bistatic FMCW RADAR principle in multiple chipless tag identification depends on the scenario of application. The postprocessing for collision detection and tag identification is similar in both methods. In this chapter, we have explored the potentiality of FMCW RADAR technique for collision detection, range extraction, and multi-tag identification in chipless RFID systems. However, there are three main steps for resolving multi-tag collision as shown in Figure 8.7.

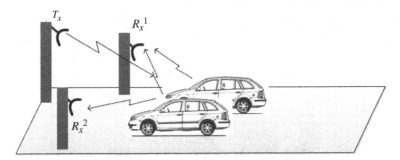

Figure 8.6 Vehicle entry system.

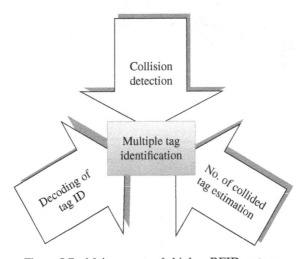

Figure 8.7 Main aspects of chipless RFID systems.

The three main steps are:

- *Step* 1: Identifying collision.
- *Step* 2: Estimation of the probable number of tags involved in collision.
- *Step* 3: Decoding of individual tag IDs from the collided response. However, sometimes it may not be possible to extract the individual tag IDs from the collided signal. But the reader can sense that a collision is there and interrogate the tags again.

A chipless RFID system with multiple tags in the interrogation zone of a reader is shown in Figure 8.8. The reader antenna sends a continuous wave (CW) signal, $x(t)$, for tag interrogation. The frequency of $x(t)$ is linearly varying with time. As described in the preceding chapter, this signal is termed as linear frequency-modulated (LFM) signal. The signal is mathematically represented by the following expression:

$$x(t) = a \cos\left[2\pi \left(f_1 t + \frac{K}{2} t^2 \right) \right] \tag{8.4}$$

Here, a is the constant amplitude of the signal over all frequency range, f_1 is the starting frequency, and K is the rate of change of frequency denoted by the following expression:

$$K = \frac{B}{T_R} \tag{8.5}$$

where B is the total bandwidth of the LFM signal and T_R is the total duration of the interrogation signal. The frequency variation of $x(t)$ is expressed mathematically as

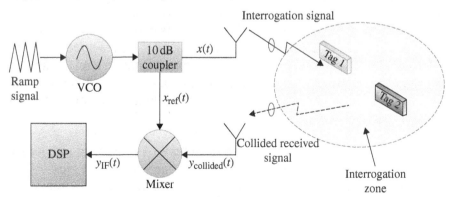

Figure 8.8 Multi-tag chipless RFID systems.

$$f(t) = f_1 + K \times t \tag{8.6}$$

On receiving the signal, the tags in the interrogation zone instantly reflect back a portion of the interrogation signal with their IDs embedded within the frequency spectrum. Let's assume that $y_1(t)$ and $y_2(t)$ are the response signals from Tags 1 and 2, respectively. Now, the response signals will be delayed by the amount of the RTOF for each tag. Tags 1 and 2 are at a distance of r_1 and r_2, respectively, from the reader antennas. Let's assume that the RTOF of Tags 1 and 2 are τ_1 and τ_2, respectively. Now, h_1 and h_2 are the transfer function of Tags 1 and 2, respectively, that cause the amplitude and phase variation in the response signals $y_1(t)$ and $y_2(t)$. The time-domain response of individual tags can be expressed as

$$y_1(t) = h_1(x_1) \tag{8.7}$$
$$y_2(t) = h_2(x_2) \tag{8.8}$$

where $x_1 = x(t - \tau_1)$ and $x_2 = x(t - \tau_2)$. The effect of random noise is not considered in this analysis. However, as long as the signal level is more than the noise floor, the methodology is valid. As the tags are in close proximity to each other, their individual response signals overlap in the time domain, and a collided signal is received by the reader antenna from the interrogation zone. It is denoted as

$$y_{\text{collided}}(t) = y_1(t) + y_2(t) = h_1(x_1) + h_2(x_2) = h_1(x(t - \tau_1)) + h_2(x(t - \tau_2)) \tag{8.9}$$

Now, the multi-tag scenario is explained with appropriate mathematical expressions, and we have the collided time-domain signal. The next step is to detect the collision, range extraction, and individual tag identification. The block diagram of the detection method using FMCW RADAR technique is shown in Figure 8.8.

A portion of the interrogation signal $x(t)$ is coupled by a 10 dB coupler and brought to a mixer. The coupled signal is represented by $x_{\text{ref}}(t)$. The signal $y_{\text{collided}}(t)$ is mixed with the reference signal in a mixer, and an IF signal is generated, which has been explained earlier. Now, by analyzing this IF signal, the collision, range extraction, and identification of chipless tags are done.

8.2.2.1 Collision Detection/Multi-Tag Detection   As described previously, for simplicity, we are considering a reading zone with two chipless RFID tags at distances r_1 and r_2, respectively. Let's assume at time instant t, the VCO generates a CW signal $x_{f1}(t)$ of frequency $f(t) = f_1$ and the signal is sent to the interrogation zone by the reader's transmitting antenna. Multiple backscattered signals are received from the tags in the interrogation zone. By the time the signal from Tag 1 reaches the reader's receiving antenna, the transmitted signal's frequency is changed. Let's assume that after a delay τ_1, $y_{f1}(t)$ is received from Tag 1. At $(t+\tau_1)$, VCO generates the signal $x_{f\tau_1}(t)$ whose frequency $f(t+\tau_1) = f_1 + (K \times \tau_1)$. When these two signals at different frequencies are mixed within the mixer, an IF signal is generated whose frequency is denoted as f_{b1}. This signal is also termed as beat frequency signal. A similar scenario occurs for the response from Tag 2. The signal from Tag 2 is received after a delay of τ_2. At time instant $(t+\tau_2)$, the VCO generates $x_{f\tau_2}(t)$ whose frequency $f(t+\tau_2) = f_1 + (K \times \tau_2)$. So after mixing, it creates an IF signal of frequency f_{b2}. The IF signals are expressed as

$$f_{b1} = f(t+\tau_1) - f(t) \tag{8.10}$$
$$f_{b2} = f(t+\tau_2) - f(t) \tag{8.11}$$

Here, we have assumed that $r_1 \neq r_2$. Therefore, τ_1 and τ_2 are different, which result in different values for f_{b1} and f_{b2}. Therefore, the mixer output contains a summation of *beat frequency* signals each corresponding to a particular tag at a particular distance within the reading zone. So the frequency content of the mixer output is expressed as

$$f_{\text{IF, combined}} = f_{b1} + f_{b2} \tag{8.12}$$

However, if there are more than two tags separated by unique distances from the reader, the IF signal will contain multiple *beat frequency* signals each corresponding to one particular tag. For N tags at distances as $r_1, r_2, \ldots, r_N$ from reader, the IF signal is represented as

$$f_{\text{IF}} = f_{b1} + f_{b2} + \cdots + f_{bN} \tag{8.13}$$

The numbers of tags that can be detected depend on the intertag distances and also the maximum reading range of the reader. As shown in Figure 8.9, five tags are in the reading zone. However, Tag 4 is beyond the maximum range of the reader, whereas Tag 5 is not that far, within the range of the reader, but does not fall within the radiation pattern of the antenna. Therefore, Tags 4 and 5 are not detected by the reader here. By analyzing the spectrum of the mixer output, the presence of multiple tags, hence collision can be detected as shown in Figure 8.10. As each *beat frequency* signal corresponds to one tag, provided that the tags are

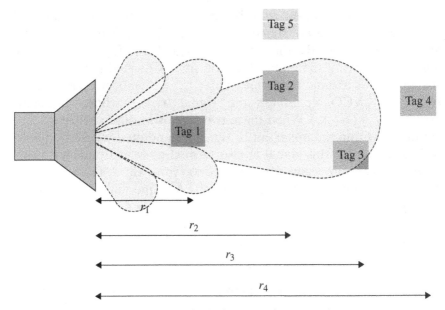

Figure 8.9 Multiple tags in the interrogation zone of an antenna.

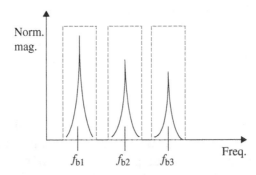

Figure 8.10 Spectrum of the IF signal containing multiple *beat frequency* signals each corresponding to a particular tag.

well separated from each other, from the number of *beat frequency* signal, the number of tags within the reading zone can be detected.

8.2.2.2 *Range Extraction of Individual Tag* From the spectrum of the IF signal, the number of *beat frequency* signals provides the information about the number of tags responding from the interrogation. The added advantage of FMCW RADAR technique is that it also provides the information of range of each tag [2, 10, 14]. For range information extraction, the amount of delay each response signal have suffered is required. Knowing this delay, the range of each tag can be calculated using expression (8.1). Again, the range can also be calculated from directly using expression (8.3).

Therefore, after extracting each beat frequency, the ranges of the tags can be determined. Now, the concern here is the minimum range resolution of the chipless RFID systems. This determines that what is the minimum distance between two tags is required for their successful detection. According to FMCW RADAR technique, the minimum range resolution is expressed as

$$\partial r = \frac{c}{2 \times B} \qquad (8.14)$$

Therefore, according to Equation 8.14, the minimum resolvable distance depends on the bandwidth of the system. For chipless RFID systems, the UWB band 3.1–10.7 GHz is used [15, 16]. It provides us with a bandwidth of 7.6 GHz, which is theoretically capable of detecting two tags that are 1.97 cm apart. Therefore, this method has the potential and viability to be used as a collision detection method for many commercial applications of chipless RFID systems. Examples include conveyer belt baggage handling, tracking of objects in indoor environment, event management, and warehouse management.

8.2.2.3 Decoding of Individual Tag IDs Along with collision detection and range information, identification of individual tag IDs is important in any RFID systems. For decoding individual tag IDs, first, we need to have an insight of the changes that tag ID causes within the IF signal. Let's first consider only one tag. The response of Tag 1 is $y_1(t)$ as expressed in Equation 8.7. Now, $h_1(x(t - \tau_1))$ is the function that possesses amplitude and phase information of the backscattered signal. Therefore, the response of Tag 1 can be expressed further as

$$
\begin{aligned}
y_1(t) &= h_1\left(x\left(t - \tau_1\right)\right) \\
&= a_1 \cos\left[2\pi\left\{f_0\left(t - \tau_1\right) + \frac{K}{2}\left(t - \tau_1\right)^2\right\} + \Psi\left(f\left(t - \tau_1\right)\right)\right] \quad (8.15) \\
&= a_1 \cos\theta_1
\end{aligned}
$$

where a_1 and θ_1 are the amplitude and phase response of Tag 1, respectively. In the mixer, $y_1(t)$ is mixed with $x_{ref}(t)$. The mixer output is

$$x_{\text{mixer}}(t) = a\cos\theta \times a_1\cos\theta_1$$

$$= \frac{aa_1}{2}\left[\cos(\theta+\theta_1)+\cos(\theta-\theta_1)\right]$$

$$= \frac{aa_1}{2}\cos\left[2\pi\left\{2f_1t+\frac{K}{2}\left(t^2+(t-\tau_1)^2\right)-f_1\tau_1\right\}+\Psi_1\left(f(t-\tau_1)\right)\right]$$

$$+ \frac{aa_1}{2}\cos\left[2\pi\left\{\frac{K}{2}\left(t^2-(t-\tau_1)^2\right)+f_1\tau_1\right\}-\Psi_1\left(f(t-\tau_1)\right)\right]$$

$$(8.16)$$

In the above equation, the first term $(aa_1/2)\cos[2\pi\{2f_0t+\frac{K}{2}(t^2+(t-\tau_1)^2)-f_0\tau\}+\psi_1(f(t-\tau_1))]$ denotes an LFM signal centered at twice the center frequency of the signal at $2f_0$. This signal is discarded by using a filter after the mixer. The second term $(aa_1/2)\cos[2\pi\{(K/2)(t^2-(t-\tau_1)^2)+f_1\tau_1\}-\psi_1(f(t-\tau_1))]$ is the beat frequency signal. Here, $\psi_1(f(t-\tau_1))$ contains the phase variation caused by resonances from the tag. The amplitude variation of the beat frequency signal can also be used for identifying the resonances [5]. Similarly, for Tag 2, we get another beat frequency signal, which can be analyzed as described previously for amplitude and phase information of the tag.

8.3 SYSTEM DESCRIPTION

The aforementioned theory is validated with a simulation setup in Agilent Advanced Design System (ADS) 2009 platform. The extracted data from the practical simulation setup in ADS 2009 is processed further in MATLAB 2010. The following section presents the system description in ADS 2009 and postprocessing in MATLAB 2010. The block diagram of the total simulation is presented in Figure 8.11. As can be seen in the figure, the system comprises of two main blocks—tags and a reader. The detailed description of the system modeling and MATLAB postprocessing are described in the following section.

8.3.1 ADS Simulation Environment

In ADS 2009, the chipless RFID tag is realized with two antennas and a series of band-stop filters [15]. Two antennas have been used with the band-stop filters, one for receiving the interrogation signal from the

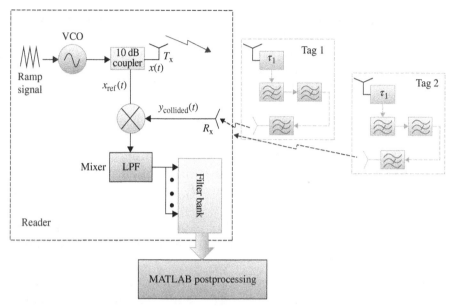

Figure 8.11   Block diagram of ADS simulation setup and postprocessing in MATLAB.

reader antenna and the other one for retransmitting the signal after modulation with the band-stop filters. The antennas are assumed to have unity gain for simplicity. Here, the presence and absence of a resonator (attenuation of power at the resonance frequency) is decoded as logic "1" and "0," respectively. To create a spatial separation and hence a delay between the response signals of two tags, two delay units (τ_1 and τ_2) are used. A power splitter splits the incoming power into two equal sections to feed to two tags. In the reader block, an inbuilt source is used for generating the interrogation chirp signal. The reader's transmitting part contains VCO driven by a ramp signal and a unity gain antenna. pilot signal is extracted from the transmission path with a 10 dB coupler. The receiving part contains an antenna, mixer, and a bank of narrowband filters. A low-pass filter is used before the filter bank to remove the high-frequency chirp signal from the mixer output. The filter bank is used to separate each *beat frequency* signal from the combined one.

As already discussed, for multiple tags identification in chipless RFID systems collision detection and the number of tags involved in collision are the two prior information need to be known before individual tag identification. As shown in Figure 8.11, a bank of narrowband band-pass filters is used to extract each *beat frequency* signal from the combined IF signal. The concern is that if the *beat frequency* signal does not fall within the passband of any of the band-pass filters then we may loss the signal from the tag.

Figure 8.12 shows the bank of narrowband filters containing n number of band-pass filters. Their passband center frequencies are f_{01}, f_{02}, ..., f_{0n}, respectively. Here, we need to consider the following system aspects:

- *Maximum reading distance*—This gives the estimation of the maximum *beat frequency* signal expected from the interrogation zone considering the tag at the maximum distance. This estimation determines the highest frequency f_{0n} for the bank of band-pass filters.
- *Minimum reading distance*—The expected minimum distance of the tag in the interrogation zone. This determines the minimum frequency f_{01} for the filter bank.
- *Range resolution*—The center frequencies of the band-pass filters are adjusted in a way to cover the entire interrogation zone. For this purpose, an approximation of expected inter-tag distance is considered in priori. For this purpose, we can use the minimum resolvable distance between two tags for the system. If the tags are

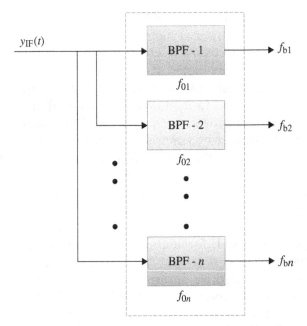

Figure 8.12 Block diagram of the bank of narrowband band-pass filter.

at a distance lower than the minimum range resolution, the system is unable to separate them. Let's assume two tags are at ∂r distance apart from each other. The difference between the beat frequency signals is calculated by the following expression (extracted using expression (8.3)):

$$\partial f = f_{b2} - f_{b1} = \frac{2 \times B}{c \times T_R}(r_2 - r_1)$$

$$= \left(\frac{2 \times B}{c}\right)\left(\frac{r_2 - r_1}{T_R}\right) \qquad (8.17)$$

$$= \frac{1}{\partial r}\left(\frac{r_2 - r_1}{T_R}\right)$$

Here, it is assumed that $r_2 - r_1 = \partial r$. Therefore, expression (8.17) becomes

$$\partial f = \frac{1}{\partial r}\left(\frac{\partial r}{T_R}\right) \qquad (8.18)$$

Therefore, the difference between adjacent center frequencies of the filters in the filter bank depends on T_R, which is the duration of the interrogation signal. The center frequency of each filter in the filter bank is calculated by the expression

$$f_{0n} = f_{01} + n \times \partial f, \quad n = 2, 3, \ldots, (n-1) \qquad (8.19)$$

The bandwidth of the filters needs to be equal to or less that ∂f as well. Otherwise, tags that are separated by ∂r may fall within the same frequency bin and cannot be separated by the method. Therefore, the frequencies and bandwidths of the filters in the filter bank needs to be adjusted carefully. The filter bank can also be implemented using FFT processing as shown in Figure 8.13 [17]. The detailed analysis of the FFT processing is described in the results section.

8.3.2   Postprocessing in MATLAB

The postprocessing on the received signal that includes separation of the collided signals and extraction of tag IDs is performed in MATLAB through signal processing. The flowchart of the overall process is shown

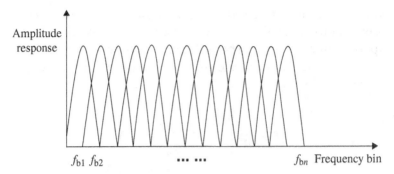

Figure 8.13 Frequency response of filter bank.

in Figure 8.14. The reader sends a CW-LFM signal to the tag. The tags respond back to the reader. The mixer generates the IF signal. Instead of using a filter bank, FFT processing is used. The spectrum of the time-domain IF signal is calculated in MATLAB through FFT. The frequency resolution of FFT is estimated using the expression

$$\partial f = \frac{F_s}{\text{NFFT}} \tag{8.20}$$

Here, F_s = sampling frequency and NFFT is the FFT block length. The signal was recorded for 150 ns duration with a sampling frequency of 1 GHz. The sampled signal is imported to MATLAB for further processing. The length of the signal is 150 points. The closely matched FFT length is 256. Therefore, a 256-point FFT was used to calculate the spectrum of the signal. Each frequency bin is 3.9 MHz. Therefore, if two signals having a frequency difference of less than 3.9 MHz, we are not being able to discriminate them from the spectrum with these calculation parameters. This resolution can be improved by recording the data for longer time duration.

Moreover, different windowing need to be used to reduce the side lobe levels of the FFT processing so that proximity tags can also be identified. The overall interrogation and postprocessing method is summarized in the flowchart shown in Figure 8.14. The process starts by sending the interrogation signal to the interrogation zone. On receiving the signal, tags residing in the interrogation zone back-scatters the signal with IDs embedded as resonances within the backscattered signal. Reader's receiving antenna receives the collided signal. The signal is then downconverted in the mixer by mixing with the reference signal. The IF signal is then processed in MATLAB for collision detection, range extraction, and tag ID decoding.

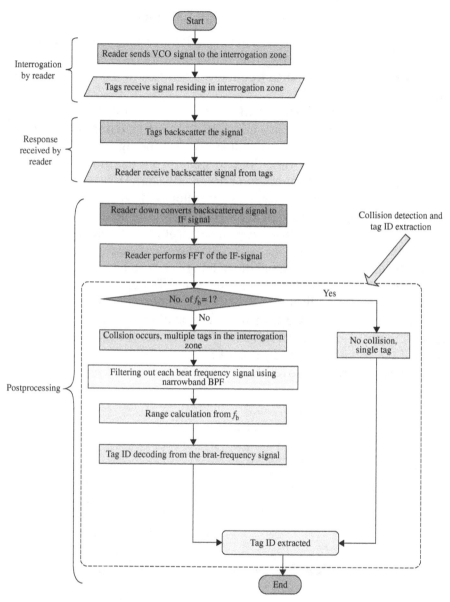

Figure 8.14 Flowchart for MATLAB postprocessing.

8.4 RESULTS AND DISCUSSION

Simulations to verify the presented theory are carried out with the described system in the preceding section. Different set of simulations are carried out for the validation of the methodology. The results of the simulation are presented in this section.

8.4.1 Collision Detection and Range Extraction

In an initial simulation, two tags having a delay of 4 ns, which resembles a physical distance of 60 cm between two tags, are considered as shown in Figure 8.15. The resonance frequencies and IDs of the tags are given in the following Table 8.1.

The tags are excited with an LFM signal, and their collided signal is mixed and downconverted in a mixer. Now, the IF signal should contain two beat frequency signals as two tags are responding.

Figure 8.16 shows the time-domain signal from the mixer. From the time-domain signal, we cannot infer about collision. However, the time-domain IF signal is transformed to frequency domain through FFT. Figure 8.17 shows the IF signal in the frequency domain. It clearly shows the presence of two *beat frequency* signals from two chipless tags in the interrogation zone. Therefore, we are having a collided response signal from which the response signal of each tag needs to be extracted.

In another simulation, two tags, Tag 1 and Tag 2, have been considered with relative distances from the reader as 7.5 and 17.5 cm, respectively. The bandwidth of the interrogation signal is 6 GHz. The time-domain IF signal is sampled at 1 GHz. Figure 8.18 shows the spectrum of the IF signal with 256-point FFT. It shows two dominant peaks in the spectrum, which proves the presence of two *beat frequency* signals.

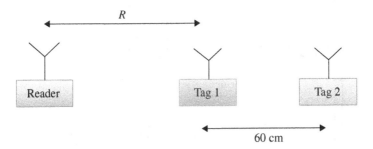

Figure 8.15 Simplified simulation setup.

Table 8.1 Specification of the Tags Used in Analysis

Tag	Resonant Frequencies (GHz)				Encoded ID
	3.5	4.1	4.25	4.75	
Tag 1	√	×	√	×	1010
Tag 2	√	×	√	√	1011

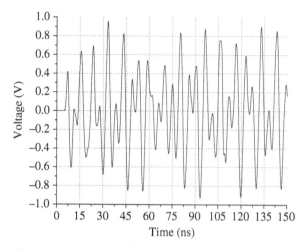

Figure 8.16 Time-domain signal $y_{IF}(t)$, output of the mixer after low-pass filtering.

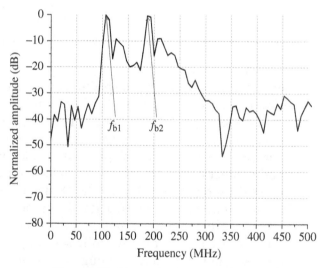

Figure 8.17 Normalized spectrum of $y_{IF}(t)$.

The difference between first and second peaks is 20 dB, and this difference is caused by the difference in path loss the two signals suffered. However, the spectrum clearly shows the presence of more than one tag, which confirms the collision. The presence of two dominant sharp peaks in the spectrum supports the conclusion that at least two tags are involved in collision. The two peaks occur at 48.83 and 70.31 MHz.

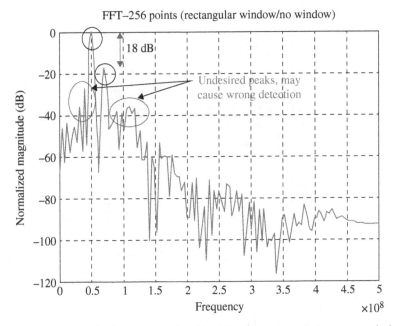

Figure 8.18 Normalized spectrum of $y_{IF}(t)$ with rectangular window or no window.

There are some undesired peaks in the spectrum, which are marked with circles. This may result wrong estimation of tag numbers as well as collision. Here, no windowing of data has been used. Using different types of windows (Hanning, Hamming, and Blackman), it is possible to reduce the side lobe level, which in turn helps to take the right decision about collision and number of tags by discarding undesired peaks.

Figure 8.19 shows the effect of windowing the IF signal data before FFT. Here, a Hanning window has been used. It reduces the sharp peaks to a much lower level (40 dB) in the first region, and in another region, it flattens the sharp peaks. Hence, by efficiently selecting a threshold value, undesired peaks can be discarded. Though windowing increases the main lobe of the two dominant peaks, it does not affect much in the range resolution as can be seen from the graph. The peaks of the blue curve (without window) and red curve (with Hanning window) remain in the same position. Therefore, windowing can be used to further improve the system's performance by reducing undesired peaks and avoiding false detection.

The ranges of the tags are calculated using expression (8.3). The range values are displayed in Figure 8.20. The ranges of the tags are 9 and 16 cm. The actual range values are 7.5 and 17.5 cm. Hence, there

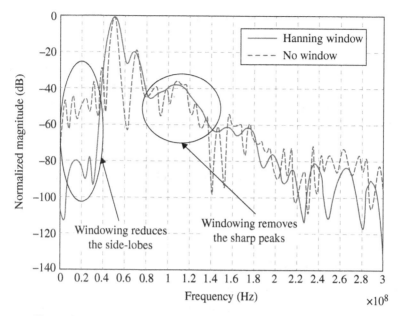

Figure 8.19   Normalized spectrum of $y_{IF}(t)$ with Hanning window.

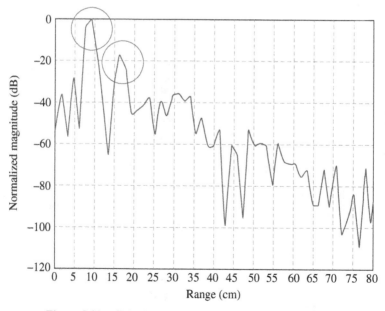

Figure 8.20   Calculated range from the spectrum of $y_{IF}(t)$.

are some range errors. This error may be reduced by the following methods:

- Recording a longer-duration data. This will increase the spectral resolution of the FFT and reduce the frequency bin.
- Calibrating the system more accurately without tag including all the system delays that may affect the actual delay as well as range.

So far, we have discussed the techniques of multiple chipless tag detection and ranging using FMCW RADAR principle. In the following section, we discuss the identification of multiple tags within the same reading zone.

8.4.2 Tag Identification

The resonances from the tags cause amplitude and phase variations or jumps in the *beat frequency* signal. Now, using the bank of band-pass filters, each beat frequency signal is singled out and analyzed for identification of tag's IDs. Figures 8.21 and 8.22 show the time-domain *beat frequency* signal $y_{b1}(t)$ and $y_{b2}(t)$, respectively. The signals have been

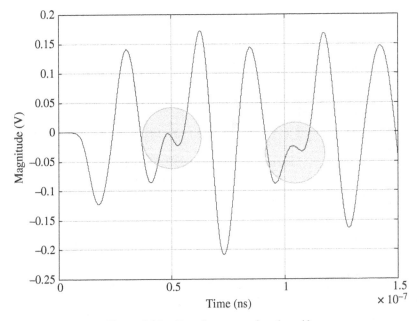

Figure 8.21   *Beat frequency* signal, $y_{b1}(t)$.

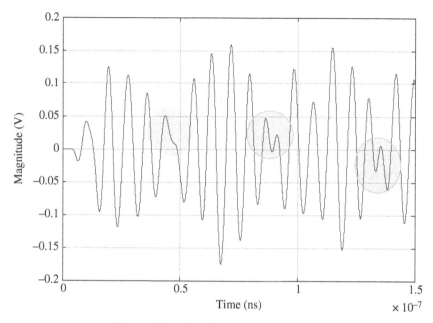

Figure 8.22   *Beat frequency* signal, $y_{b2}(t)$.

extracted from the combined IF signal $y_{IF}(t)$ by filtering out using narrowband band-pass filters. Figure 8.21 shows the time-domain signal $y_{b1}(t)$ with two shaded regions where phase reversal and amplitude jumps have occurred. Similarly, in Figure 8.22, three shaded regions are shown. These are because of the resonances from the tags. Now, to detect the resonances, a method has been proposed called as Hilbert transform (HT) assisted complex analytical signal representation [5, 14, 18–20]. It has also been explained in detail in Chapter 4. However, for convenience, a brief of the method is explained here.

The time-domain signal $y_{b1}(t)$ and $y_{b2}(t)$ is a real-time signal. To have the envelope and phase information, the signal is converted to an analytic signal to extract the envelope and phase information [19, 21]:

$$\text{Magnitude variation} = \left| \text{HT}\big(y_b(t)\big) \right| \tag{8.21}$$

$$\text{Phase variation} = \arg\left[\text{HT}\big(y_b(t)\big) \right] \tag{8.22}$$

The aforementioned transformation method is applied for each of the extracted *beat frequency* signal. The resonance information is extracted from the magnitude and phase variation in the *beat frequency* signal. However, in this method, we get the resonance information with time. We need to find out which resonance corresponds to which frequency. Figure 8.23 shows the time–frequency relationship of the VCO signal. So the time of sending each frequency is known in priori. In Figure 8.23, the rectangular regions are the regions of expected resonances from the tags. Therefore, 3.5 GHz resonance is expected around 40 ns and so on. So by comparing the time information of resonances, the resonance frequency (ID bits) can be identified. Another advantage is that any undesired resonance other than the marked region can also be discarded.

Figure 8.24 shows the recovered magnitude response from Tag 1 (1010). The lower x-axis shows the time of resonance and the upper x-axis shows the frequency of the resonance. Figure 8.25 shows the first, third, and fourth resonances of Tag 2 (1011). There is another magnitude variation around 4.4 GHz. We may cross validate the resonance information from phase values as whether we can discard or accept it as a resonance.

Figure 8.26 shows the phase variation obtained from Tag 1. It matches with the magnitude variation showing phase variations at first and

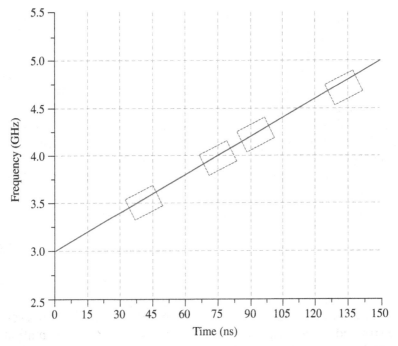

Figure 8.23 Time–frequency relationship of $x(t)$, expected resonance positions is highlighted.

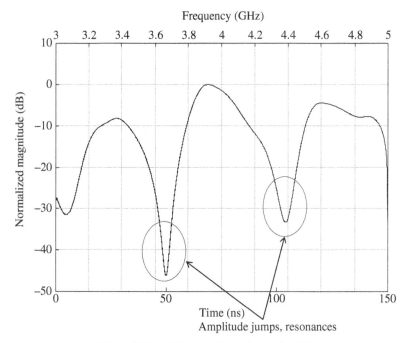

Figure 8.24 Calculated envelope of $y_{b1}(t)$.

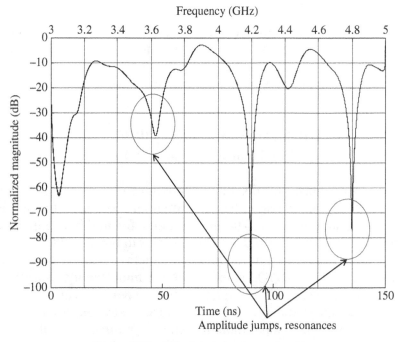

Figure 8.25 Calculated envelope of $y_{b2}(t)$.

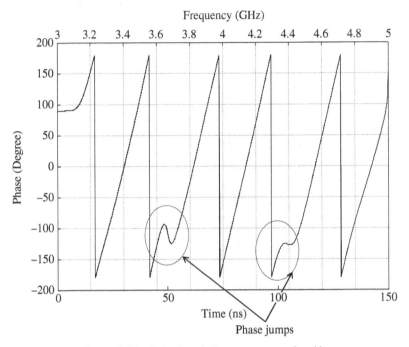

Figure 8.26   Calculated phase response of $y_{b1}(t)$.

third resonance positions. However, in the case of Tag 2 as shown in Figure 8.27, the phase variations are seen in the first, third, and fourth resonance positions. Therefore, we can discard the amplitude variation at 4.4 GHz as an undesired magnitude variation. Therefore, the decoded tag IDs are:

- Tag 1: first and third resonances, ID—"1010"
- Tag 2: first, third, and fourth resonances, ID—"1011"

So from the above results and analysis, it can be concluded that the FMCW RADAR technique has the potential to be used for multiple chipless tag identification. The methodology for multiple chipless tag identification is summarized with key aspects in Figure 8.28.

The results presented in the preceding section validate the potentiality of the FMCW RADAR technique for multi-tag identification in chipless RFID systems. However, the collision detection and ranging of the tags residing in the interrogation zone are extracted from the spectrum of the IF signal. Separation of each *beat frequency* signal from the collided IF signal can be done by either using a filter bank or through

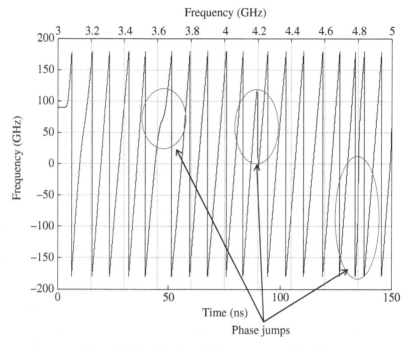

Figure 8.27 Calculated phase response of $y_{b2}(t)$.

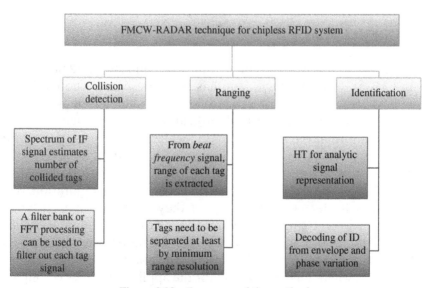

Figure 8.28 Summary of the method.

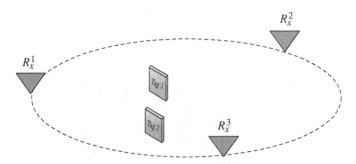

Figure 8.29 Multiantenna system, multistatic RADAR.

FFT processing. Here, the results have been presented for FFT processing. As long as the tags have an intertag distance more than the minimum range resolution of the system, the proposed method is capable of detecting collision and ranging. For extracting individual tag IDs, the beat frequency signals are analyzed further through HT.

However, if the tags are from same distances from the reader, the FMCW RADAR technique alone may not be able to detect the collision; in such scenario, beamforming will be advantageous. The use of multiple receivers is also advantageous. As shown in Figure 8.29, for R_x^1, the tags are at same distances. But from R_x^2 and R_x^3, the tags are at different distances. Therefore, the IF signal from R_x^2 and R_x^3 will infer the collision but not the IF signal from R_x^1.

8.5 CONCLUSION

In this chapter, a novel method of multiple chipless tag identification using FMCW RADAR technique has been presented. The method is capable of detecting the presence of multiple chipless tags together with range information and identification of individual tag IDs. The range information further helps in localizing of chipless tags within the interrogation zone. This will enable localization together with identification within a much lower cost than the conventional RFID localization. However, for monostatic RADAR configuration, this method fails to detect multiple tags if they reside at same distances from the receiver. Using multiple receivers or beamforming will solve the problem. This method together with beamforming will provide better performance for multi-tag identification in chipless RFID systems. This will open new horizon for the application areas of chipless RFID systems.

REFERENCES

1. R. E. Azim and N. Karmakar, "*A collision avoidance methodology for chip-less RFID tags*," in *Microwave Conference Proceedings (APMC), 2011 Asia-Pacific*, Melbourne, VIC, December, 5–8, 2011, pp. 1514–1517.

2. M. I. Skolink, *Introduction to Radar Systems*, New York: McGraw-Hill, 1962.

3. K. Penttila, M. Keskilammi, L. Sydanheimo, and M. Kivikoski, "Radar cross-section analysis for passive RFID systems," *IEE Proceedings Microwaves, Antennas and Propagation*, vol. 153, pp. 103–109, 2006.

4. I. Balbin and D. N. Karmakar, "*Radio Frequency Transponder System*," Australian Provisional Patent, DCC, Ref:30684143/DBW, October 20, 2008.

5. R. V. Koswatta and N. C. Karmakar, "*A novel method of reading multi-resonator based chipless RFID tags using an UWB chirp signal*," in *Microwave Conference Proceedings (APMC), 2011 Asia-Pacific*, Melbourne, VIC, December 5–8, 2011, pp. 1506–1509.

6. M. Brandl, S. Schuster, S. Scheiblhofer, and A. Stelzer, "*A new anti-collision method for SAW tags using linear block codes*," in *IEEE International Frequency Control Symposium, 2008*, Honolulu, HI, May 19–21, 2008, pp. 284–289.

7. S. Scheiblhofer, S. Schuster, and A. Stelzer, "Signal model and linearization for nonlinear chirps in FMCW Radar SAW-ID tag request," *IEEE Transactions on Microwave Theory and Techniques*, vol. 54, pp. 1477–1483, 2006.

8. H. Eugin and L. Jong-Hun, "*A method for multi-target range and velocity detection in automotive FMCW radar*," in *12th International IEEE Conference on Intelligent Transportation Systems, 2009. ITSC '09*, St. Louis, MO, October 4–7, 2009, pp. 1–5.

9. A. E. Carr, L. G. Cuthbert, and A. D. Olver, "Digital signal processing for target detection FMCW radar," *IEE Proceedings F Communications, Radar and Signal Processing*, vol. 128, pp. 331–336, 1981.

10. K. Pourvoyeur, R. Feger, S. Schuster, A. Stelzer, and L. Maurer, "*Ramp sequence analysis to resolve multi target scenarios for a 77-GHz FMCW radar sensor*," in *11th International Conference on Information Fusion, 2008*, Cologne, June 30–July 3, 2008, pp. 1–7.

11. Available at http://www.slideshare.net/tobiasotto/principle-of-fmcw-radars (accessed on July 27, 2012).

12. Available at http://www.radartutorial.eu/05.bistatic/bs04.en.html (accessed on September 4, 2012).

13. T. Axness, R. V. Coffman, B. A. Kopp, and K. W. O'Haver, "Shared aperture technology development," *Johns Hopkins APL Technical Digest*, vol. 17, pp. 285, 1996.

14. S. Mukherjee, "*Chipless radio frequency identification by remote measurement of complex impedance,*" in *European Microwave Conference,* Munich, October 9–12, 2007, pp. 1007–1010.

15. S. Preradovic, I. Balbin, N. C. Karmakar, and G. F. Swiegers, "Multiresonator-based chipless RFID system for low-cost item tracking," *IEEE Transactions on Microwave Theory and Techniques,* vol. 57, pp. 1411–1419, 2009.

16. S. Preradovic and N. C. Karmakar, "Chipless RFID: bar code of the future," *IEEE Microwave Magazine,* vol. 11, pp. 87–97, 2010.

17. G. Brooker, *Introduction to Sensors,* Raleigh: SciTech Publishing, 2008.

18. D. R. Brunfeldt and S. Mukherjee, "*A novel technique for vector network measurement,*" in *37th ARFTG Conference Digest-Spring,* Boston, MA, June, 1991, pp. 35–42.

19. J. Detlefsen, A. Dallinger, S. Schelkshorn, and S. Bertl, "*UWB millimeter-wave FMCW radar using Hubert transform methods,*" in *IEEE Ninth International Symposium on Spread Spectrum Techniques and Applications,* Manaus-Amazon, August 28–31, 2006, pp. 46–48.

20. S. Mukherjee, "Remote Characterization of Microwave Networks — Principles and Applications," in *Advanced Microwave Circuits and Systems,* V. Zhurbenko, Ed., Shanghai: In-Tech, 2010.

21. C. Capus, Y. Rzhanov, and L. Linnett, "The analysis of multiple linear chirp signals," *IEE Colloquium (Digest),* vol. 19, pp. 49–55, 2000.

CHAPTER 9

CHIPLESS TAG LOCALIZATION

9.1 INTRODUCTION

Radio-frequency identification (RFID) is a wireless data communication technology. The generic RFID system has two main components: (i) an RFID tag and (ii) a reader. Occasionally, the reader is connected to the central database for continuous updating of information. A block diagram of a typical RFID system is shown in Figure 9.1. Tags contain application-specific integrated circuits (ASICs) where different modulation techniques are used for data encoding. The reader sends an RF signal toward the tags, and the tags reply back with their IDs to the reader.

In conventional chip-based RFID systems, the main cost for the tag comes from the ASIC [1]. The cost of the tag is the main hindrance for low-cost item tagging in massive volume. To address this issue, significant research is undergoing for the development of low-cost chipless RFID tags. Chipless RFID systems offer key benefits over conventional RFID systems such as low-cost and on-demand printability on papers or polymers with conventional printing methods and conductive inks, which make it apt for mass deployment in low-cost item tagging for a wide

Chipless Radio Frequency Identification Reader Signal Processing, First Edition.
Nemai Chandra Karmakar, Prasanna Kalansuriya, Rubayet E. Azim and Randika Koswatta.
© 2016 John Wiley & Sons, Inc. Published 2016 by John Wiley & Sons, Inc.

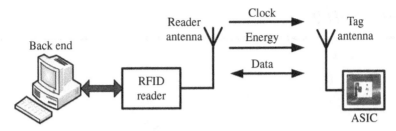

Figure 9.1 Conventional chip-based RFID system.

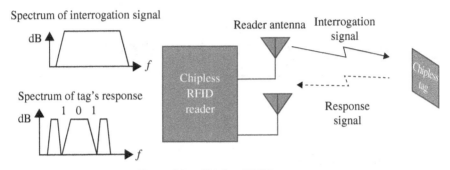

Figure 9.2 Chipless RFID system.

range of applications [2]. A number of chipless tags based on time-domain (TD) [3], frequency-domain (FD) [4], phase-based [5], and imaging-based techniques [6] have been reported in the open literature. However, the FD tags reported in Refs. [7–9] have higher data capacity than their predecessors [3]. Figure 9.2 shows the basic operating method of a FD chipless RFID system [7–9]. The chipless tag is interrogated using an ultra-wideband (UWB) signal. The tag ID is extracted from the presence and absence of resonances at predefined frequencies in the received signal [7].

9.2 SIGNIFICANCE OF LOCALIZATION

Till date, the research on chipless RFID focuses on tag design and identification [3–12]. But in addition to tag identification, knowing the location of a tag is very important for many RFID applications. For example, in asset tracking, the location of an asset is an important aspect. Localization of tagged items is also important in automatic health monitoring, vehicle entry, and access control. Localization of tag can be particularly important in multitag environment where one particular tag needs to be addressed within a cluster of tags [13]. Tag localization methods [13–15] presented for conventional chipped RFID system rely on the processing capability of onboard chips [14] and therefore cannot be used for chipless RFID tags. The only reported work on localization of chipless RFID tags is for surface acoustic wave (SAW) tags [16, 17]. But due to the presence of interdigital transducer (IDT) and piezoelectric substrate, SAW tags are costly, rigid, and not readily printable on paper or polymer. Instead, FD chipless RFID tags are showing enormous potential in terms of low-cost, bit-capacity, and on-demand printability with moderate reading range. These advantageous features make them competent for large-scale deployment for low-cost item tagging. However, to the authors' best knowledge, the printable FD chipless RFID tags have not been analyzed for localization purpose as yet.

This chapter proposes a novel localization technique for printable FD chipless RFID tags. The localization is done by analyzing the TD backscattered signal of the tag by UWB-IR ranging technique. A short-duration UWB-IR signal is used as an interrogation signal [18]. A set of multiple transceivers are placed at fixed positions in the zone of interrogation. Chipless tags are placed at multiple positions inside the zone. The transmitters illuminate the tag with UWB-IR signals. The collocated receivers receive the backscatter signal from the tag. The round-trip time of flight (RTOF) is calculated from the *structural mode response* of the backscattered signal from the tag. From the RTOF, the relative distances between the tag and the receivers are estimated. Subsequently, the tag position in the interrogation zone is determined using linear least squares (LLS) approximation [19]. First, the accuracy of the method is analyzed by placing tags at different known positions. Afterward, arbitrary unknown tag positions are estimated with the localization method with known error.

9.3 TAG LOCALIZATION: CHIPLESS VERSUS CONVENTIONAL RFID

As stated earlier, localization of tags has significant importance in RFID systems. It improves the process automation and tracking of objects. This section starts with a comprehensive overview of the localization methods used in conventional RFID systems. Their limitations while applied for chipless RFID systems are discussed next. At the end the review, importance and significance of localization in chipless RFID systems are presented.

9.4   CONVENTIONAL RFID TAG LOCALIZATION TECHNIQUES

Locating of an object in the interrogation zone involves two main tasks: (i) ranging of the object from known receivers and (ii) locating the position of the object in the interrogation zone. However, based on the ranging procedure used for any particular type of tag, the localization can be divided into three main categories. They are:

i. RTOF estimation [17, 20]
ii. Received signal strength (RSS)-based localization [21]
iii. Phase evaluation method [20]

Detailed description of the mentioned methods is given below.

9.4.1   RTOF Estimation

For RTOF estimation, the reader transmits the interrogation signal and waits for replies from the tags. The total time that the signal requires to travel from the reader to the tag and from the tag to the reader is called RTOF. From the measured RTOF, the distance of the tag from a particular receiver is calculated. Figure 9.3 shows a conventional RFID tag interrogation with a reader. The timing is illustrated in Figure 9.4. The transceiver sends the interrogation signal at time t_{TX}. The tag retransmits the modulated signal after a processing period of t_{tag}. The time of flight (TOF) to the tag and back to the reader is denoted as t_1 and t_2. On knowing the time instances t_{TX} and t_{RX}, the distance of the tag is calculated as

$$d_{tag} = \frac{t_1 + t_2}{2} \cdot c = \frac{(t_{RX} - t_{TX}) - t_{tag}}{2} \cdot c \qquad (9.1)$$

Figure 9.3   Conventional RFID interrogation.

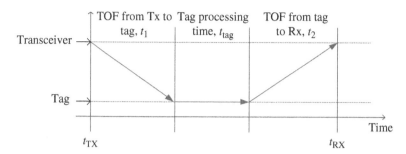

Figure 9.4 RTOF measurement.

The tag processing delay needs to know exactly in priori; otherwise, it will cause measurement error. However, the quality of measurement depends on the signal bandwidth, time synchronization between the transmitted signal and the received echo, and signal-to-noise ratio (SNR) [22]. As the range/distance of tag from the transceiver is known, afterward, the location of the tag can be determined by trilateration algorithm [17] using known approximation methods or using directional antennas with beamforming capabilities.

However, in estimating the accuracy of delay, Cramer and Rao suggested a lower bound based on bandwidth and SNR in the received signal [23]. It is called as Cramér–Rao lower bound (CRLB). The CRLB indicates the lower bound of the delay error by the following expression [24]:

$$\sigma_{\hat{\tau}}^2 \geq \frac{1}{8\pi^2 (\text{BW})^2 \text{SNR}} \tag{9.2}$$

Here, $\sigma_{\hat{\tau}}^2$ is the error of delay estimation, and BW is the bandwidth of the received signal. The CRLB for the ranging distance can be obtained as $c.\sigma_{\hat{\tau}}$ where c is the speed of light/EM wave (3×10^8 m/s).

Figure 9.3 shows the CRLBs for ranging errors in terms of SNR for four bandwidths. It is seen that for BW = 0.75 GHz, the theoretical lower bound varies for 1.5 cm up to 14 cm for SNR value of 10 to −10 dB. For higher bandwidth such as for 6 GHz bandwidth, the error remains between 0.2 and 1.8 cm for different values of SNR. Therefore, in this case, the dependency on SNR is less than the previous case. In our analysis, we will be using a 6 GHz UWB signal for measuring the RTOF, which should give good accuracy in acceptable range according to CRLB (Fig. 9.5).

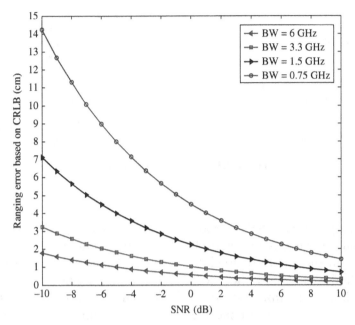

Figure 9.5 CRLB for ranging errors with UWB pulse under different SNR conditions.

9.4.2 RSS-Based Localization

In RSS-based method, the tag-reader distance versus RSS is mapped in priori. On receiving a tag's response, the reader can match with mapped RSS for estimating its range. SpotOn [14] and LANDMARC [14] are two well-known methods where RSS is used to estimate the tag position. In SpotOn, the tag is modified so that the RF signal attenuation can be utilized to provide intertag distance. In LANDMARC [14], multiple reference tags are used to estimate the position of the targeted tag. This can also be done with a moving reader or a rotating antenna [14, 20, 21]. However, the accuracy is affected by multipath and it requires on-site adaptation [13].

9.4.3 Phase Evaluation Method

In phase evaluation method, the round-trip phase is measured to estimate the distance of the tag. It requires a two-way communication between the tag and the reader where the tag is capable of generating phase coherent response. An array antenna can also be used for the angle-of-arrival (AOA) measurement [25]. However, the method of localization depends largely on the type of transponders such as conventional active, passive, and semipassive tags and/chipless RFID tags.

The localization techniques for conventional RFID systems rely on the processing capability of the onboard chip of the RFID tag. Due to the absence of an onboard IC, the methods are not straight away applicable for chipless RFID tag localization. Therefore, novel and dedicated localization methods are required for chipless RFID tags. The main aim of the work is to fill up the gap in research.

9.5 CHIPLESS RFID TAG LOCALIZATION

The contrasting and challenging issues in chipless tag localization are that (i) the chipless tags are unable to modify the backscattered power according to the requirement for distance estimation in RSS method due to the absence of ASIC, (ii) they are unable to establish intertag communication with nearby reference tags, (iii) a large number of reference points may be required for adaptation of RSS-based method in chipless RFID tag localization, (iv) the phase-based method is not useful as the tags are unable to provide phase coherent response, and finally, (v) the tags are incapable of sending any beacon signal to the reader for localization purpose. Thus, for localization and identification purposes, the reader solely depends on the backscattered signal from the tag. This left RTOF-based technique as a potential approach for chipless RFID tag localization. In time-domain reflectometry (TDR)-based SAW chipless tags, tag localization has been analyzed through RTOF measurement method [16]. The SAW tag is interrogated with the time inverse of its impulse response, and the tag responds back the autocorrelated signal. The maximum of the autocorrelated peak gives the RTOF estimation [16, 17]. In Ref. [17], a localization accuracy of ±20 cm has been reported using three receiving antennas and SAW tags. Arumugam et al. [16] have reported a subdecimeter localization accuracy (3.17 cm) for SAW tags. In Ref. 26, a unipolar monopole antenna has been studied for localization purpose. But it mainly focuses on the identification rather than tag localization. To the author's best knowledge, the FD chipless tags reported in Refs. 1, 5, 9–11, 27, 28 have not yet been analyzed for ranging and localization purposes.

The FD chipless tags can be interrogated either using a linearly frequency-modulated continuous-wave (LFMCW)/chirp signal [12, 29] or with UWB-IR signal [18, 28, 30, 31]. Due to the ultrashort duration of the pulse, such UWB signals can provide high-resolution localization

[26, 30, 32]. In this chapter, the chipless tags are interrogated with short-duration UWB pulse, and the backscattered signal from the tag is analyzed for extracting the range information from the structural mode response. The method is presented in Section 9.7.

9.6 BENEFITS OF CHIPLESS TAG LOCALIZATION

The importance of localization with low-cost RFID tags is huge [20]. With diverse application areas, the ability to localize chipless tags will benefit in many ways. Among them, some are listed here:

Reliable tag reading: Reliable reading of chipless tags is challenged by factors such as (i) undesired reflections from nearby objects, (ii) interferences from proximity tags, and (iii) interference from surrounding environment. After locating the positions of each tag using the proposed method, the reader's antenna beam can be precisely toward the tags. This enables the reader to acquire more precise responses from tags with minimum interference from nearby objects or proximity tags. Thus, it improves the reading reliability significantly.

Multiple tag reading: Multiple tag reading is a challenging issue in chipless RFID systems. Due to the absence of onboard ICs, none of the established anticollision or communication protocols can be implemented in chipless RFID systems. Therefore, the multitag reading is fully a task of the reader, without any supportive communication from the tag. On knowing the accurate location of each tag present in the interrogation zone, the reader can steer its antenna beam toward tags one by one and thereby read multiple tags.

Tracking of tagged items: Tracking of tagged items is particularly important in healthcare monitoring, supply chain management, and mining industry. Accurate localization of chipless tags will lower the deployment cost for automation and tracking of low-cost items in these applications, which is otherwise not commercially feasible with chipped RFID tags. It will help process automation without much human involvement.

Automatic entry control: Tracking and detection of objects or persons equipped with chipless RFID tags enable automatic responses like door opening, turning on lights, or triggering of alarms.

So far, comprehensive analysis on tag localization, research gap identification, and significance of chipless tag localization has been presented. In the next section, the proposed method is presented.

9.7 PROPOSED LOCALIZATION FOR CHIPLESS RFID TAGS

This section describes a novel localization method for FD chipless RFID tags. For the proposed method, the RTOF information is extracted from the backscattered signal of the tag. Therefore, analyzing and understanding the backscattered response signal from chipless tag is very important here. As there is no onboard IC on the tags, the properties of backscattered signal need to be used for localization. Therefore, in this section, the backscattered signal from chipless tag has been analyzed and represented with a unique signal model including Gaussian noise present in the reading zone. Then, the ranging and positioning techniques are explained in detail.

9.7.1 Backscattered Signal from Chipless Tag

In this analysis, a multibit slot-loaded chipless RFID tag [9] as shown in Figure 9.6a is used. The FD response of the tag is shown in Figure 9.6b. It has four resonances at 6.6, 7.2, 7.9, and 9 GHz. The tag is interrogated by a UWB-IR pulse, which is expressed as

$$x(t) = A\cos(2\pi f_c t)\exp\left(-\frac{(t-\tau)^2}{2\sigma^2}\right)\mathbb{S} \qquad (9.3)$$

Here, f_c is the center frequency of the pulse, τ is the time index for the peak value of the pulse, and σ is the variance. When interrogated with

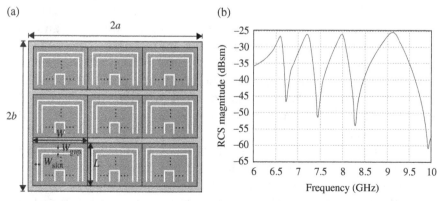

Figure 9.6 (a) Layout of the chipless tag, $L = 6.4$ mm, $W = 8.4$ mm, $W_{\text{slot}} = 0.2$ mm, and $W_{\text{gap}} = 0.2$ mm, on Taconic TLX – 8.0. (b) Backscattered RCS of 4-bit tag (simulation).

such a pulse signal, the backscattered signal from the tag has two scattering modes: (i) structural mode response $y_s(t)$ and (ii) tag mode response $y_t(t)$.

The initial backscattering from the tag, $y_s(t)$, depends on the size of the tag and is very similar to the interrogation pulse. The delayed backscattering is the tag mode, $y_{tag}(t)$, which contains the resonance information [18, 33]. Hence, the overall TD RCS response from the tag is expressed as

$$
\begin{aligned}
y_{tag}(t) &= y_s(t) + y_t(t) + n(t) \\
&= A_s \cos(2\pi f_c t) \exp\left[-\frac{\left(t - \tau - \tau_{tag}\right)}{2\sigma^2} \right] \\
&\quad + \sum_{n=1}^{N} A_n \exp\left(a_n + j\omega_n t\right) + n(t)
\end{aligned}
\tag{9.4}
$$

Here, $n(t)$ is the Gaussian random noise. Here, the tag's resonance information is represented as the summation of exponentially decaying signals, and the number of exponentials depends on the number of resonances present in the tag. A_n is the complex amplitudes and $(a_n + j\omega_n t)$ are poles corresponding to predefined resonance frequencies [31, 34]. τ_{tag} is the delay involved in the tag's backscattering and depends on the distance between the tag and the transceiver. The interrogation pulse and the backscattered responses in TD from the tag are shown in Figure 9.7a and b. As we can see in Figure 9.7a and b, the interrogation pulse is centered at 1.2 ns, whereas the structural mode response from the tag is centered at 2 ns. Therefore, $\tau_{tag} = 0.8$ ns represents the RTOF for the tag. Using TD rectangular windowing techniques on the tag's backscattering shown in Figure 9.7b, two modes are separated, and their frequency contents are analyzed for tag information. The frequency analysis of the separated responses (structural and tag modes) is shown in Figure 9.7c. As discussed, the structural mode contains the similar frequency content as the interrogation pulse, whereas the late-time tag mode contains the resonance information of the chipless tag. In this case, for RTOF = 0.8 ns yields the tag's range of 12 cm.

The TD structural RCS signal $y_s(t)$ has the maximum magnitude within the overall RCS from the tag as can be seen in Figure 3.2b. The time delay τ between the interrogation pulse $x(t)$ and the structural mode RCS $y_s(t)$ is inversely proportional to fourth order of distance between the tag and the receiver. This is called the RTOF for a particular

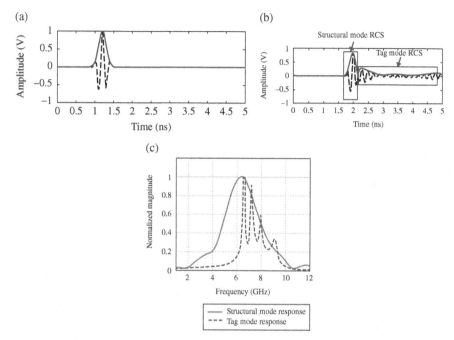

Figure 9.7 (a) UWB-IR pulse for interrogation, (b) time-domain (TD) backscattering from tag, and (c) normalized magnitude spectrum of the structural mode and tag mode response.

tag. The RTOF τ is identified within the overall RCS response $y_{tag}(t)$ through peak detection algorithm [35]. From τ, the range information of the tag is extracted and used to localize the exact position of it in the interrogation zone.

9.7.2 Maximum Detection Range

The maximum detection range of the tag is described by the well-known *radar* equation [25]. It is expressed as

$$r_{t,max}^{4} = \frac{G_a^2 P_t \lambda^2 \sigma_s}{(4\pi)^3 P_{r,min}} \tag{9.5}$$

Here, G_a is the gain of the transmitting and receiving antennas, P_t denotes transmitted power, λ is the wavelength of the center frequency of the operating bandwidth, σ_s is structural mode RCS of the chipless tag, $P_{r,min}$ is the minimum detectable power of the receiver, and $r_{t,max}$ is the maximum detectable range of the chipless tag. In our setup with vector network analyzer (VNA) and horn antennas, the noise level is measured as −60 dB. In order to successfully detect and decode the tag ID, it has been found from the measurement that the backscattered RCS from the tag need to be 10 dB greater than the noise level. Therefore, $P_{r,min}$ is chosen −50 dBm. The center of the operating frequency band is 9 GHz, which gives $\lambda = 3.33$ cm. The transmitting and receiving antennas used here have a gain of $G_a \approx 11$ dBi. The VNA transmits −15 to 3 dBm power over the entire frequency band. Since we use an array of multiple slot-loaded square patches, for simplicity, we consider the tag as a rectangular flat plate and approximate the structural mode RCS σ_s from physical optics according to the following expression [36]:

$$\sigma_s = \frac{64\pi a^2 b^2}{\lambda^2} \cos^2 \varphi \left[\frac{\sin(2ka\sin\varphi)}{2ka\sin\varphi} \right] \tag{9.6}$$

Here, $k = 2\pi/\lambda$, $2a$, and $2b$ are two sides of the chipless tag, and φ is the aspect angle. The chipless tag has a dimension of 6.5 cm × 4.5 cm, which gives a structural mode RCS $\sigma_s \approx -11$ dBsm at 9 GHz frequency according to expression (9.6).

Figure 9.8 shows the dependency of maximum range of the tag $r_{t,max}$ on transmitted power and backscattered structural mode RCS σ_s of the

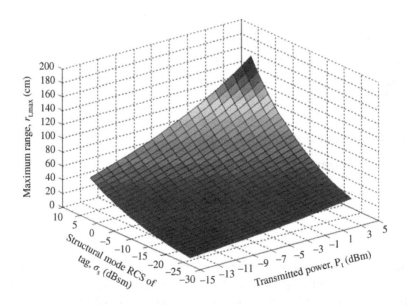

Figure 9.8 Dependency of maximum range of the tag $r_{t,max}$ on transmitted power and backscattered structural mode RCS σ_s.

tag. For our given setup, with the chipless tag described here, a maximum of 80 cm detection range can be obtained. Thus, we consider two separate setups. In the first setup, we placed the transceivers on a circle of 70 cm diameter, which gives an observation area of $70 \times 70 \, \text{cm}^2$ where $r_{t,max} = 20 \, \text{cm}$ is considered ($P_t = -15 \, \text{dBm}$). In the second setup, the transceivers are placed on the perimeter of a circle of 120 cm, which allows an observation area of $1.2 \times 1.2 \, \text{m}^2$ where $r_{t,max} = 55 \, \text{cm}$ ($P_t = 0 \, \text{dBm}$). The SNR also plays an important role in the accuracy of the range estimation according to CRLB. The CRLB indicates the lower bound of the range estimation error by the following expression [24]:

$$\text{Delay estimation error}, c.e_\tau \geq \frac{c}{\left\{ 8\pi^2 B^2 \left(\text{SNR} \right) \right\}^{1/2}} \tag{9.7}$$

Here, B is the bandwidth that is 6 GHz in our setup, and c is the speed of EM wave (3×10^8 m/s). Figure 9.9 shows the CRLBs for ranging errors in terms of SNR for different bandwidths. As the bandwidth increases, the ranging error decreases, and for a particular bandwidth, it improves with the improvement of SNR. However, for 6 GHz bandwidth, the theoretical lower bound of ranging error varies from 0.2 to 1.8 cm for different SNRs.

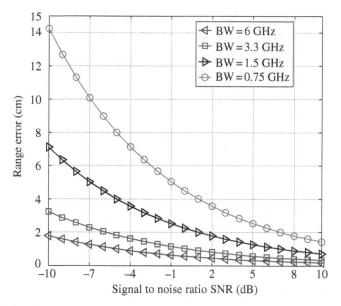

Figure 9.9 CRLBs for ranging errors in terms of SNR for different bandwidths.

9.7.3   Localization of Tag

For unambiguous localization in a 2D plane, a minimum of three transceivers is required. Figure 9.10 illustrates the analysis setup and coordinate system for chipless tag localization. The transceivers are placed at
equal radial distance R from the centroid of the interrogation zone
(here, a square room is considered; hence, the centroid is the intersection point of the diagonals). The angular separation between adjacent
transceivers is expressed as

$$\Phi_{\mathrm{d}} = \frac{2\pi}{m} \tag{9.8}$$

Here, m is the number of transceivers. If three transceivers are used,
then the angular separation between adjacent transceivers is 120°
according to expression (9.8). When a tag enters in the interrogation zone,
the backscattered signal from the tag is captured by the transceivers

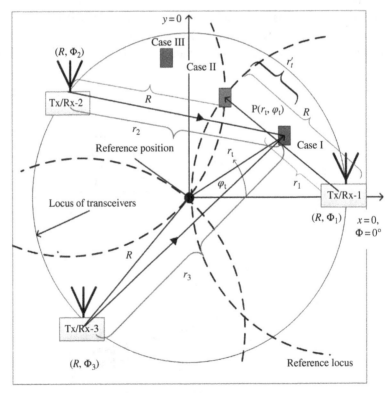

Figure 9.10   Coordinate system and transceiver location for trilateration for chipless
RFID systems.

placed around the interrogation zone and analyzed for the time of arrival for the structural mode to calculate the RTOF and hence the range of the tag. After the relative ranges from multiple transceivers are calculated, the position of the tag is estimated using LLS approximation.

We are using three transceivers in our analysis. Ideally, using the respective ranges with the known positions of the transceivers, the intersection of three circles around the transceivers should give the exact position of the chipless tag as shown in Figure 9.11a. The three circles intersect at a single point, which is the location of the tag.

A real challenge is to locate the tag when the range measurement is erroneous. Then, the circles do not converge to a single point but rather create a common overlapping region as shown in Figure 9.11b. The shaded region ABC is found by trilateration. In this situation, the LLS approximation is used for finding the exact location of the tag with minimal error. The details of the ranging and LLS method are explained later in this section.

For estimating the estimation error of the method for localizing the chipless tag, first, the tag is placed at various known positions denoted by $P(r_t, \varphi_t)$, and the position is estimated using the localization method. Afterward, it is compared with the actual tag position to estimate the estimation error. The tag is placed at six different angular positions (30°, 90°, 150°, 240°, 270°, and 330°) with respect to $\Phi = 0°$ for each of the four radial distances (5, 10, 15, and 20 cm) from the center of the interrogation zone.

The localization starts by sending an UWB-IR signal for interrogation. The chipless tag backscatters the signal, which is received by the receiving antennas. The localization technique for chipless tag is broadly

(a)   (b)

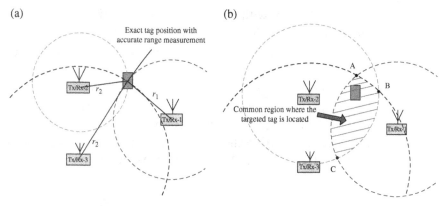

Figure 9.11 (a) Trilateration by intersection of three circles with error-free range measurement and (b) trilateration by intersection of three circles with erroneous range measurement.

divided into two parts: *ranging* and *positioning* of the tag. During *ranging*, the relative distances of the tag from the receivers are calculated from the RTOF of the received signals at different receiver positions. These distances are used to estimate the location (r_t, φ_t) of the tag in the interrogation zone during *positioning*.

9.7.4   Ranging of Tag

Due to spatial differences, the signal from the tag to the receiving antennas suffers from measurable delays. For localization purpose, the delays between the transmitted and received signals are calculated. The response received from the tag has two parts as described earlier: (i) structural mode response $y_s(t)$ and (ii) tag mode response $y_t(t)$. It has been established in Refs. 18, 31 that the structural mode RCS has larger signal amplitude compared to the tag mode response and possesses similarity with the interrogation UWB-IR pulse. The cross-correlative peak detection method can be used by correlating the backscattered time series with the interrogation pulse shape to identify the delay of the structural mode RCS. The delay can also be calculated from the envelope of the TD backscattered response signal. As already described in the previous section, the structural RCS has the highest magnitude within the overall response. Therefore, the maximum in the envelope of the TD RCS response is used to detect the delay τ_{tag}. However, instead of using actual RTOF, we have used a relative delay and range measurement method in our analysis. The advantages are as follows:

i. We do not need to know the exact time index for the interrogation pulse enabling us to use the S-parameters captured by the VNA [37].

ii. The center of the interrogation zone can be calibrated as the reference point.

iii. A simple relative range measurement method can be applied. Here, the transceivers are fixed in their positions on the perimeter of the circle. Therefore, the center is used as a reference position. After calibrating the reference position, reference locus can be considered for each of the transceivers as shown in Figure 3.3.

Figure 9.12 depicts backscattered signal from the tag for the reference position $(r_t = 0)$ and an arbitrary position $P(r_t, \varphi_t)$. However, for

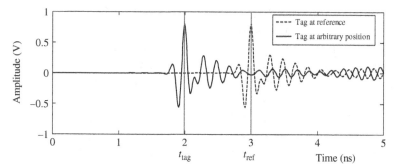

Figure 9.12   UWB-IR received signal for tags at reference position and arbitrary position.

calculating the envelope of the TD signal, the received signal $y_{tag}(t)$ is converted to an analytic signal $H(y_{tag}(t))$ by Hilbert transform [38]. Then, the peak value is extracted from the envelope of the signal [39, 40]. The time of arrival of the structural mode is given by the following expression:

$$t_{tag} = \arg\max_{t}\left\{H\left(y_{tag}(t)\right)\right\} \tag{9.9}$$

Here, t_{tag} is the time instant where the tag's backscatter signal has its maximum magnitude and equals to the value of t, which maximizes $|H(y_{tag}(t))|$. From Figure 3.5, the times of arrivals of the structural mode (maximum in the envelope of the backscattered signal) for the reference and tag signals are t_{ref} and t_{tag}, respectively. Therefore, the ranges of the tag can be calculated from the following expressions:

$$r' = \left|t_{tag} - t_{ref}\right| \times \frac{c}{2} \tag{9.10}$$

$$r = \begin{cases} R + r', & \text{for } t_{tag} > t_{ref} \text{ (Case I)} \\ R, & \text{for } t_{tag} = t_{ref} \text{ (Case II)} \\ R - r', & \text{for } t_{tag} < t_{ref} \text{ (Case III)} \end{cases} \tag{9.11}$$

9.7.5   Positioning of Tag

As already described, the range measurement has some errors, which is also depicted from CRLB. Therefore, the tag position cannot be estimated in a straightforward way by finding the intersection point of three circles. Therefore, LLS approximation is applied to estimate the tag position with minimal least squares error.

The position of the tag can be determined once the distance from known receivers is estimated. The solution is actually finding the intersection points of several circumferences. If m-number of receivers are used, the tag position is estimated by solving m nonlinear equations. For estimating the tag position, the *LLS*, singular value decomposition (SVD) [37], or nonlinear least squares (*Newton*) method [37, 41] can be used. In this chapter, the *LLS* method is used to estimate the tag position.

Based on the tag and the antenna positions shown in Figure 3.3, the relative distance r_i between the tag at $P(r_t, \varphi_t)$ and the transceiver at (R, Φ_i) is represented by the following expression:

$$\left(r_t \cos\varphi_t - R\cos\Phi_i\right)^2 + \left(r_t \sin\varphi_t - R\sin\Phi_i\right)^2 = r_i^2$$

To avoid complexity of nonlinear solution, the system expressions are converted to a linear system by using a linearization tool [37, 42]. Here, the jth expression is used to linearize the system, and the expressions are

$$\left(r_t \cos\varphi_t - R\cos\Phi_i\right)^2 + \left(r_t \sin\varphi_t - R\sin\Phi_i\right)^2 - \left(r_t \cos\varphi_t - R\cos\Phi_j\right)^2 + \left(r_t \sin\varphi_t - R\sin\Phi_j\right)^2 = r_i^2 - r_j^2$$

Expanding and rearranging the terms of this expression lead to the following expression:

$$\left(\cos\Phi_j - \cos\Phi_i\right)r_t \cos\varphi_t + \left(\sin\Phi_j - \sin\Phi_i\right)r_t \sin\varphi_t = b_{ij} \qquad (9.12)$$

Here, b_{ij} is expressed as

$$b_{ij} = \left(\frac{1}{2D}\right)\left(r_i^2 - r_j^2\right) \qquad (9.13)$$

So for m-number of receivers, the linearized system of expressions is

$$\left(\cos\Phi_j - \cos\Phi_1\right)r_t \cos\varphi_t + \left(\sin\Phi_j - \sin\Phi_1\right)r_t \sin\varphi_t = b_{1j}$$

$$\left(\cos\Phi_j - \cos\Phi_2\right)r_t \cos\varphi_t + \left(\sin\Phi_j - \sin\Phi_2\right)r_t \sin\varphi_t = b_{2j}$$

$$\vdots$$

$$\left(\cos\Phi_j - \cos\Phi_m\right)r_t \cos\varphi_t + \left(\sin\Phi_j - \sin\Phi_m\right)r_t \sin\varphi_t = b_{mj}$$

These system equations are written in matrix form as [19, 41, 43]

$$C\vec{r} = \vec{b}. \qquad (9.14)$$

where

$$
C = \begin{bmatrix} \left(\cos\Phi_j - \cos\Phi_1\right) & \left(\sin\Phi_j - \sin\Phi_1\right) \\ \left(\cos\Phi_j - \cos\Phi_2\right) & \left(\sin\Phi_j - \sin\Phi_2\right) \\ \vdots & \vdots \\ \left(\cos\Phi_j - \cos\Phi_m\right) & \left(\sin\Phi_j - \sin\Phi_m\right) \end{bmatrix}_{(m\times2)}, \vec{b} = \begin{bmatrix} b_{1j} \\ b_{2j} \\ \vdots \\ b_{mj} \end{bmatrix}_{(m\times1)},
$$

$$
\text{and} \quad \vec{r} = \begin{bmatrix} r_t \cos\varphi_t \\ r_t \sin\varphi_t \end{bmatrix}_{(2\times1)}
$$

Here, C is known as the fixed transceivers' positions, which are known in priori. $\vec{b}$ is calculated by expression (9.13) using r_i and r_j values obtained from ranging, and $\vec{r}$ is the unknown tag position. Now, $\vec{r}$ is determined such that the mean square error for expression (9.14) has a minimum value. This condition is expressed as [44]

$$
\underset{\vec{r}\in\mathbb{R}}{\text{Min}} \| C\vec{r} - \vec{b} \|^2 \tag{9.15}
$$

Solving expression (9.14) such that it validates the condition expressed by (9.15), the unknown tag position $\vec{r}$ is determined.

9.8   RESULTS AND DISCUSSION

9.8.1   Simulation Environment

In order to obtain a backscattered signal close enough to the realistic scenario as shown in Figure 9.10, a 3D model is constructed with chipless tag described in Section 9.7, and multiple probes as the receivers in CST Microwave Studio Suite 2012 and full-wave electromagnetic simulation are performed by moving the tag at different positions within the interrogation zone. A UWB-IR signal of 6 GHz bandwidth and 1.2 ns duration is used to interrogate the tag. In our simulation and laboratory experimental setup, we have used three transceivers at three known positions in the interrogation zone. The chipless tag is moved within the interrogation zone at different positions denoted by $P(r_t, \varphi_t)$ as already described.

From the CST simulation model, the tag response signals at the probes are extracted for RTOF calculation. The RTOF data is used to estimate the relative distance between tag and receiving antennas. For positioning, the mathematical formulation of the localization problem is done in MATLAB as a system of nonlinear equations. The nonlinear system is linearized using a linearization tool to an approximated linear system. The linear system is then solved by applying LLS approximation for estimating the position of the tag.

9.8.2 Experimental Setup

The experimental validation of the proposed localization method has been carried out in Monash Microwave, Antenna, RFID, and Sensor Laboratory (MMARS) as shown in Figure 9.13. The experimental setup consists of an Agilent VNA PNA E8361A as the reader electronics with a pair of horn antennas as reader antennas and a fabricated chipless tag.

The horn antennas operate from 6 to 12 GHz with an almost flat gain of 11 dBi over the band. The tag is fabricated on Taconic TLX-0. The measurement and postprocessing for localization are presented in the flow chart shown in Figure 9.14. As can be seen from the flow chart, three measurements are taken for the tag localization purpose:

i. Background measurement
ii. Reference/calibration measurement
iii. Tag measurement

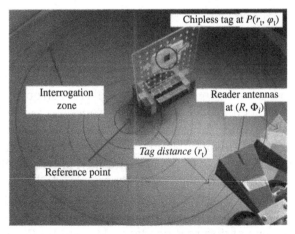

Chipless tag at $P(r_t, \varphi_t)$

Interrogation zone

Reader antennas at (R, Φ_i)

Tag distance (r_t)

Reference point

Figure 9.13 Schematic of simulation and experimental setup for localization.

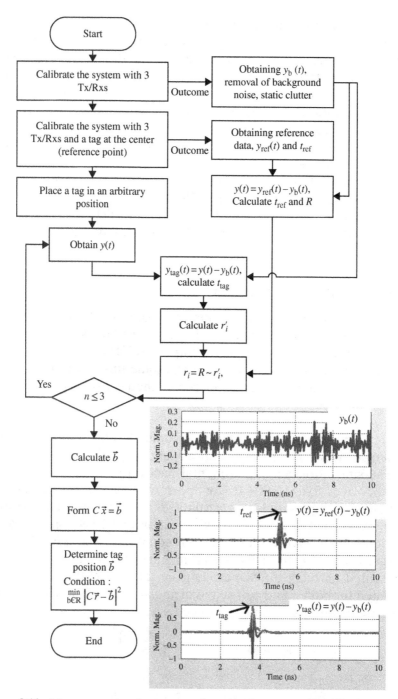

Figure 9.14   Measurement and postprocessing steps for proposed localization method.

The background measurement is performed with the receivers only, without any tag. Assuming static clutter contributions, the subtraction of background measurement from the tag measurement removes the clutter, antenna coupling, cable effects, and background noise. The reference measurement is taken to calibrate the center of the interrogation zone for relative range measurement. To remove the interference from nearby receivers and undesired reflections from walls, time gating is used. The TD backscattered RCS from a chipless tag can be directly obtained by a subnanosecond sampler. However, in a laboratory environment, we have used a VNA as the reader system. Therefore, the FD RCS data is stored in VNA and converted to TD RCS through IFFT. From the TD RCS, the time of arrival of the structural mode RCS is calculated, and the relative range r_i is estimated.

9.8.3 Results and Discussion

This section presents the simulation and experimental results for the proposed localization method. Their close resemblance validates the proposed localization technique for chipless RFID tags. First, the result of error analysis is presented; afterward, some unknown tag positions have been estimated with the proposed method considering the known error probability.

9.8.3.1 Accuracy The localization accuracy has been analyzed from the perspective of linear tag distance r_t and angular position φ_t of the tag. The simulation and experimental backscattered response signals are imported to MATLAB for postprocessing for ranging and positioning of chipless tag. For each data set, the ranging and tag position (r_t, φ_t) are estimated through peak detection and LLS approximation technique, respectively, from system equations. The deviations of measured φ_t and r_t from the actual angular and linear positions are represented through root mean square error (RMSE). RMSE is calculated for both r_t and φ_t. The radial and angular position estimation errors are denoted as e_{r_t} and e_{φ_t}, respectively:

$$e_{r_t} = (\text{MSE})^{\frac{1}{2}} = \left\{ \frac{1}{n} \sum_{i=1}^{n} (\hat{r}_t - r_t) \right\}^{\frac{1}{2}} \tag{9.16}$$

$$\varphi_{r_t} = (\text{MSE})^{\frac{1}{2}} = \left\{ \frac{1}{n} \sum_{i=1}^{n} (\hat{\varphi}_t - \varphi_t) \right\}^{\frac{1}{2}}$$

Here, $\hat{r}_t$ and $\hat{\varphi}_t$ are the mean values for r_t and φ_t. An accurate estimation of (r_t, φ_t) with a minimal of error leads to successful localization of the chipless tag in the interrogation zone.

Accuracy in Angular Position Estimation The results for angular tag position measurement are presented in Figures 9.15 and 9.16. The measured angular tag position for different linear distance of tag from origin is shown in Figure 9.15. Both the simulation and experimental results are included in the graph. The measured angular position of the tag is close enough to the actual angular positions with some exceptions. As for $r_t = 5$ and 10 cm and $\varphi_t = 30°$, the measured value differs for almost 10° from the actual tag position. This is the maximum deviation observed from simulation data. However, in experimental results, a maximum of 13° is observed for $r_t = 10$ cm and $\varphi_t = 330°$. Figure 9.16 shows the RMSE for angular position estimation, e_{ϕ_t}, from the actual angular position of the tag for different values of φ_t. For simulation, a maximum average deviation of 7° is observed for $\varphi_t = 30°$. However, in all other angular positions, the deviation remains below 3°. In actual measurement, the deviation varies from maximum 5° to 1.3°. The 5° variation occurs for $\varphi_t = 90°$, and for all other angular positions, the deviation remains below 3.5°. However, other than some discrepancies in some cases, for most of

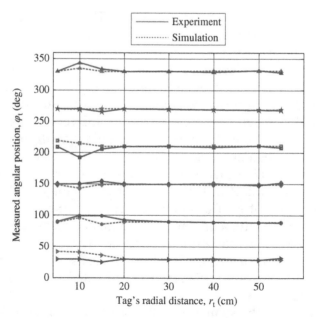

Figure 9.15 Measured angular tag position with varying linear distance.

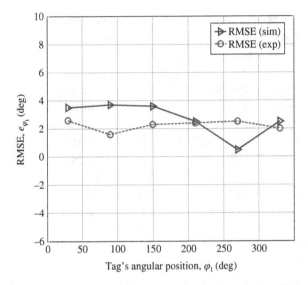

Figure 9.16 RMSE with respect to actual angular position.

the angular positions for all values of r_t, the measured angular tag position matches closely to the actual angular position.

Accuracy in Linear Tag Position Estimation The measured results of linear tag distance estimation are presented in this section. Figure 9.17 presents the measured result for linear tag distance estimation with respect to actual tag distance for different values of φ_t. The scattered points depict the measured radial distances of tags for different angular position and radial distances, whereas the solid curve shows the average measured radial distance of the tag for four different values of φ_t. For φ_t, the measured linear distance varies within 1 cm from the actual value. For some angular tag positions, as for φ_t, it deviates a maximum of 2 cm, and for all other tag positions, it remains within 1.3 cm. Figure 9.18 shows the RMSE for tag range, from the actual tag range for various values of φ_t. The simulation and experimental values are close to each other with a minimum of 0.5 cm and maximum of 2.7 cm, whereas the experimental result deviates an average of 1 cm for most values of φ_t except in two cases where it shows 3 cm deviation. On average, ~2 cm variation is observed in estimated tag linear distance from the actual position.

Table 9.1 shows the summary of the result of error analysis. In the tag's radial distance estimation, varies by 0.4 cm between simulation

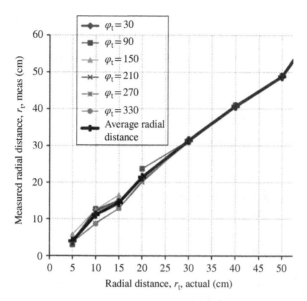

Figure 9.17 Measured linear tag position with varying angular positions.

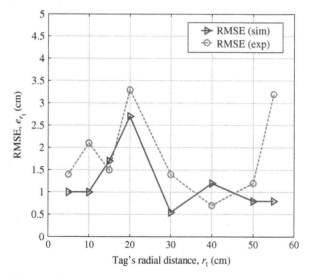

Figure 9.18 Average deviation from actual tag position.

Table 9.1   Error in Position Estimation of Chipless Tag

Error Type	Radial Error		Angular Error	
	e_{r_t} (Simulation)	e_{r_t} (Measured)	e_{φ_t} (Simulation)	e_{φ_t} (Measured)
RMSE	1.7 cm	2.1 cm	3.5°	3.5°

and measured results. However, in angular position estimation, the RMSE is the same for simulation and measured. The estimation error clearly states that the method is capable of estimating the chipless tag's position with 2 cm and 3.5° error in φ_t and respectively.

9.8.4 Unknown Tag Localization

In the first step, the estimation error for the localization method by placing tags at known positions is calculated, and the result has been presented in the previous section. Next, the method is used to localize randomly placed tags using the proposed method. First, the relative ranges are calculated, followed by application of LLS method for tag position estimation.

Figure 9.19 depicts the tags in the interrogation zone. Table 9.2 represents the result of the tag position estimation for some of the arbitrary positions. However, from the earlier analysis, we have concluded that the estimated errors in radial distance and angular position estimation are 2.1 cm and 3.5°, respectively. Therefore, the estimated tag positions vary within ±2.1 cm and ±3.5° in radial distance and angular positions,

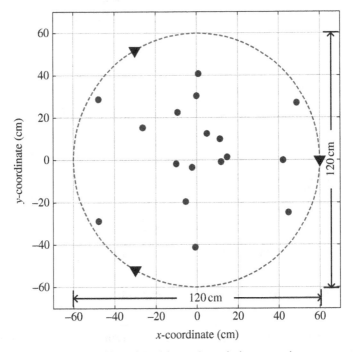

Figure 9.19 Estimated positions of tags in interrogation zone.

Table 9.2 Arbitrary Tag Position Estimation

Position	Estimated Tag Position	
	r_t (cm)	φ_t (°)
1	20.3	254
2	13.5	69
3	10.3	189
4	15	42
5	14.7	6
6	11.8	356
7	56	211
8	55.8	29
9	42	0

respectively. As shown in Table 9.2, for tag position 1, the estimated tag radial distance is 20.3 cm; considering the error, the radius can be 18.2 or 22.4 cm. Therefore, it will remain within 4 cm of linear distance. Whereas the angular position is estimated as 254°, it can vary from 250.5° to 257.5° within 6° of angular variation. Similar results are obtained for all other arbitrary positions. The proposed method yields simple but accurate localization of the chipless tag in the interrogation zone within acceptable RMS error.

9.9 CONCLUSION

A comprehensive review for both chipped and chipless RFID tag localization and an explanation of a novel method for localization of frequency signature-based chipless RFID tags have been presented in this chapter. Tag localization benefits the RFID system by improving the reliability of tag reading process, tracking of objects, and even collision detection and mitigation of collision scenario by DOA estimation using beamforming algorithm. A unique RTOF calculation for frequency signature-based chipless tag from UWB-IR interrogation signal and time-gating analysis has been presented here. A unique signal conditioning method is used for representing the TD response of chipless tag with structural and tag mode RCS together with ambient noise model. The structural mode RCS is the most dominant response and easily distinguishable, and the delay of this response compared to the interrogation pulse gives the RTOF and relative range of the tag. The tag mode RCS yields the ID information. A TD windowing for both of the RCS modes yields precise range and ID information. This finding is unique in chipless RFID tag localization and has not been reported before. A set of linear equations have been developed based on the trilateration localization method and has been applied for tag position and RMSE estimation. The theory is validated through extensive simulation both in EM solver CST and in-house developed MATLAB codes and experimental setup. In the measurement setup, a set of calibration processes were developed to subtract background noise and obtain the reference data for tag reference and RTOF reference. Based on the calibrated information, the relative ranges of the tag for three transceivers and position data were extracted. This unique set of calibration not only removes the ambient random noise but also converge to accurate determination of range and position (r_t, φ_t). The analysis, synthesis, and application of extracted data reveal very accurate, reliable tag localization for six valid situations of investigation for 5–20 cm read range and 30–330° angular resolution. The analysis based on RMSE revalidated the accuracy with max ±2.1 cm and ±3.5° of resolution. Finally, arbitrary tag position investigation has yielded very accurate results.

The adaptation of the proposed backscattered signal-based localization method is applicable to different types of FD chipless tag and does not require any modification within the tag structure. Therefore, it can be employed with existing chipless RFID systems with the postprocessing for localization implemented in the back-end IT layer. There are some challenges that may affect the performance of the method. In a multitag scenario, response from multiple tags may overlap and makes it challenging to calculate the RTOF for any of the tags. This will affect the accuracy of the localization method. A high sampling rate is required for capturing the short-duration, high-resolution response signals from the tag.

REFERENCES

1. S. Preradovic and N. C. Karmakar, "Chipless RFID: bar code of the future," *IEEE Microwave Magazine*, vol. 11, pp. 87–97, 2010.

2. R.-E. A. Anee, S. M. Roy, N. C. Karmakar, R. Yerramilli, and G. F. Swiegers, "Printing Techniques and Performance of Chipless Tag Design on Flexible Low-Cost Thin-Film Substrates," in *Chipless and Conventional Radio Frequency Identification: Systems for Ubiquitous Tagging*, N. C. Karmakar, ed.: Hershey, PA: IGI Global, 2012, pp. 175–195.

3. A. Ramos, D. Girbau, A. Lazaro, and S. Rima, "*IR-UWB radar system and tag design for time-coded chipless RFID,*" in *6th European Conference on Antennas and Propagation (EUCAP)*, Prague, March 26–30, 2012, pp. 2491–2494.

4. S. Preradovic, I. Balbin, N. C. Karmakar, and G. F. Swiegers, "Multiresonator-based chipless RFID system for low-cost item tracking," *IEEE Transactions on Microwave Theory and Techniques*, vol. 57, pp. 1411–1419, 2009.

5. I. Balbin and N. C. Karmakar, "Phase-encoded chipless RFID transponder for large-scale low-cost applications," *IEEE Microwave and Wireless Components Letters*, vol. 19, pp. 509–511, 2009.

6. R. Das (2006, 9 Nov, 2012). *Chip-Less RFID—The End Game*, Cambridge, MA: IDTechEx. Available at http://www.idtechex.com/products/en/articles/00000435.asp (accessed July 4, 2015).

7. S. Preradovic and N. C. Karmakar, "*Design of fully printable planar chipless RFID transponder with 35-bit data capacity*," in *European Microwave Conference, 2009. EuMC 2009*, Rome, September 29–October 1, 2009, pp. 013–016.

8. I. Balbin and N. Karmakar, "*Radio Frequency Transponder System*," Australian Provisional Patent, DCC, Ref:30684143/DBW, October 20, 2008.

9. M. A. Islam and N. C. Karmakar, "A novel compact printable dual-polarized chipless RFID system," *IEEE Transactions on Microwave Theory and Techniques*, vol. 60, pp. 2142–2151, 2012.

10. M. S. Bhuiyan, R. Azim, and N. Karmakar, "*A novel frequency reused based ID generation circuit for chipless RFID applications*," in *2011 Asia-Pacific Microwave Conference Proceedings (APMC)*, Melbourne, VIC, December 5–8, 2011, pp. 1470–1473.

11. M. A. Islam and N. Karmakar, "*Design of a 16-bit ultra-low cost fully printable slot-loaded dual-polarized chipless RFID tag*," in *2011 Asia-Pacific Microwave Conference Proceedings (APMC)*, Melbourne, VIC, December 5–8, 2011, pp. 1482–1485.

12. S. Preradovic and N. C. Karmakar, "*Multiresonator based chipless RFID tag and dedicated RFID reader*," in *Microwave Symposium Digest (MTT), 2010 IEEE MTT-S International*, Anaheim, CA, May 23–28, 2010, pp. 1520–1523.

13. M. Bouet and A. L. dos Santos, "*RFID tags: Positioning principles and localization techniques*," in *Wireless Days, 2008. WD '08. 1st IFIP*, Dubai, November 24–27, 2008, pp. 1–5.

14. T. Sanpechuda and L. Kovavisaruch, "*A review of RFID localization: Applications and techniques*," in *5th International Conference on Electrical Engineering/Electronics, Computer, Telecommunications and Information Technology, 2008. ECTI-CON 2008*, 2008, pp. 769–772.

15. J. Zhou and J. Shi, "RFID localization algorithms and applications—a review," *Journal of Intelligent Manufacturing*, vol. 20, pp. 695–707, 2009.

16. D. D. Arumugam, V. Ambravaneswaran, A. Modi, and D. W. Engels, "2D localisation using SAW-based RFID systems: a single antenna approach," *International Journal of Radio Frequency Identification Technology and Applications*, vol. 1, pp. 417–438, 2007.

17. T. F. Bechteler and H. Yenigun, "2-D localization and identification based on SAW ID-tags at 2.5 GHz," *IEEE Transactions on Microwave Theory and Techniques*, vol. 51, pp. 1584–1590, 2003.

18. P. Kalansuriya and N. Karmakar, "*Time domain analysis of a backscattering frequency signature based chipless RFID tag,*" in *Asia-Pacific Microwave Conference Proceedings (APMC)*, Melbourne, VIC, December 5–8, 2011, pp. 183–186.

19. X. Xiaochun, S. Sahni, and N. S. V. Rao, "*On basic properties of localization using distance-difference measurements,*" in *11th International Conference on Information Fusion*, Cologne, June 30–July 3, 2008, pp. 1–8.

20. R. Miesen, R. Ebelt, F. Kirsch, T. Schafer, L. Gang, W. Haowei, and M. Vossiek, "Where is the Tag?," *IEEE Microwave Magazine*, vol. 12, pp. S49–S63, 2011.

21. C. Alippi, D. Cogliati, and G. Vanini, "*A statistical approach to localize passive RFIDs,*" in *Proceedings of the IEEE International Symposium on Circuits and Systems (ISCAS)*, Island of Kos, Greece, May 21–24, 2006, pp. 843–846.

22. L. Gang, D. Arnitz, R. Ebelt, U. Muehlmann, K. Witrisal, and M. Vossiek, "*Bandwidth dependence of CW ranging to UHF RFID tags in severe multipath environments,*" in *2011 IEEE International Conference on RFID (RFID)*, Orlando, FL, April 12–14, 2011, pp. 19–25.

23. H. Urkowitz, *Signal Theory and Random Processes*. Dedham, MA: Artech House, 1983.

24. C. Woo Cheol and H. Dong-Sam, "*An accurate ultra wideband (UWB) ranging for precision asset location,*" in *2003 IEEE Conference on Ultra Wideband Systems and Technologies*, 2003, pp. 389–393.

25. M. I. Skolink, *Radar Handbook*, 3rd Ed. New York: McGraw-Hill 2008.

26. H. Sanming, Z. Yuan, L. Choi Look, and D. Wenbin, "Study of a uniplanar monopole antenna for passive chipless UWB-RFID localization system," *IEEE Transactions on Antennas and Propagation*, vol. 58, pp. 271–278, 2010.

27. T. Singh, S. Tedjini, E. Perret, and A. Vena, "*A frequency signature based method for the RF identification of letters,*" in *2011 IEEE International Conference on RFID (RFID)*, Orlando, FL, April 12–14, 2011, pp. 1–5.

28. A. Vena, E. Perret, and S. Tedjini, "High-capacity chipless RFID tag insensitive to the polarization," *IEEE Transactions on Antennas and Propagation*, vol. 60, pp. 4509–4515, 2012.

29. R. V. Koswatta and N. C. Karmakar, "*A novel method of reading multi-resonator based chipless RFID tags using an UWB chirp signal,*" in *Microwave Conference Proceedings (APMC), 2011 Asia-Pacific*, Melbourne, VIC, December 5–8, 2011, pp. 1506–1509.

30. A. Vena, E. Perret, and S. Tedjni, *"Novel compact RFID chipless tag,"* in *Progress in Electromagnetics Research Symposium, PIERS 2011 Marrakesh,* Marrakesh, Morocco, March 20–March 23, 2011, pp. 1062–1066.

31. P. Kalansuriya and N. Karmakar, *"UWB-IR based detection for frequency-spectra based chipless RFID,"* in *Microwave Symposium Digest (MTT), 2012 IEEE MTT-S International,* Montreal, QC, June 17–22, 2012, pp. 1–3.

32. W. Shaohua, Z. Qinyu, F. Rongfei, and Z. Naitong, *"Match-filtering based TOA estimation for IR-UWB ranging systems,"* in *International Wireless Communications and Mobile Computing Conference, 2008. IWCMC '08,* Crete Island, August 6–8, 2008, pp. 1099–1105.

33. S. Jang, W. Choi, T. K. Sarkar, M. Salazar-Palma, K. Kyungjung, and C. E. Baum, "Exploiting early time response using the fractional Fourier transform for analyzing transient radar returns," *IEEE Transactions on Antennas and Propagation,* vol. 52, pp. 3109–3121, 2004.

34. M. Manteghi, *"A novel approach to improve noise reduction in the matrix pencil algorithm for chipless RFID tag detection,"* in *2010 IEEE International Symposium on Antennas and Propagation and CNC-USNC/URSI Radio Science Meeting: Leading the Wave, AP-S/URSI 2010,* Toronto, ON, Canada, July 11–July 17, 2010.

35. P. Kalansuriya, N. C. Karmakar, and E. Viterbo, "On the detection of frequency-spectra-based chipless RFID using UWB impulsed interrogation," *IEEE Transactions on Microwave Theory and Techniques,* vol. 60, pp. 4187–4197, 2012.

36. R. A. Ross, "Radar cross section of rectangular flat plates as a function of aspect angle," *IEEE Transactions on Antennas and Propagation,* vol. 14, pp. 329–335, 1966.

37. H. Karl (2005, 10 Jul 2012). *Protocols and Architectures for Wireless Sensor Networks.* Hoboken, NJ: Wiley.

38. R. V. Koswatta and N. C. Karmakar, "A novel reader architecture based on UWB chirp signal interrogation for multiresonator-based chipless RFID tag reading," *IEEE Transactions on Microwave Theory and Techniques,* vol. 60, pp. 2925–2933, 2012.

39. F. Blanchard, L. Razzari, H. C. Bandulet, G. Sharma, R. Morandotti, J. C. Kieffer, T. Ozaki, M. Reid, H. F. Tiedje, H. K. Haugen, and F. A. Hegmann, "Generation of 1.5 µJ single-cycle terahertz pulses by optical rectification from a large aperture ZnTe crystal," *Optical Express,* vol. 15, pp. 13212–13220. 2007.

40. Y. Depeng, A. E. Fathy, L. Husheng, M. Mahfouz, and G. D. Peterson, *"Millimeter accuracy UWB positioning system using sequential sub-sampler and time difference estimation algorithm,"* in *2010 IEEE Radio and Wireless Symposium (RWS),* New Orleans, LA, January 10–14, 2010, pp. 539–542.

41. F. Izquierdo, M. Ciurana, F. Barcelo, J. Paradells, and E. Zola, "*Performance evaluation of a TOA-based trilateration method to locate terminals in WLAN,*" in *2006 1st International Symposium on Wireless Pervasive Computing*, IEEE, January 16–18, 2006, pp. 1–6.

42. W. Murphy and W. Hereman (1995, 10 Jul 2012). *Determination of a Position in Three Dimensions Using Trilateration and Approximate Distances*. [tech. report]. Available at http://inside.mines.edu/~whereman/papers/Murphy-Hereman-Trilateration-MCS-07-1995.pdf (accessed July 4, 2015).

43. C. Dionisio, "*Detection and localization of lost objects by SAR technique,*" in *IGARSS '96. International Geoscience and Remote Sensing Symposium, Remote Sensing for a Sustainable Future*, Lincoln, NE, May 27–31, 1996, pp. 28–30, vol. 1.

44. F. Reichenbach, A. Born, D. Timmermann, and R. Bill, "*A distributed linear least squares method for precise localization with low complexity in wireless sensor networks,*" presented at the Proceedings of the Second IEEE international conference on Distributed Computing in Sensor Systems, San Francisco, CA, June 18–20, 2006.

CHAPTER 10

STATE-OF-THE-ART CHIPLESS RFID READER

10.1 INTRODUCTION

This chapter proposes a smart chipless RFID reader system. A smart reader refers to a stand-alone chipless RFID reader, which sense the presence of the tagged object, starts the reading process, and updates the database automatically. Therefore, the identification process is fully automated and does not require any human involvement. Figure 10.1 summarizes the contents of the book according to the chapters. The previous book by these authors has summarized the reader architecture and hardware implementation for chipless RFID system [1]. The main focus of this book is to provide readers with the developments and research directions are followed for chipless RFID reader in terms of detection, accuracy improvement, multiple tag reading problems, and advanced signal processing. Chapters 2–8 focus on improving the efficacy of the chipless RFID reader. In Chapters 2, 3, and 4, three novel techniques for detecting the resonances in the backscattered signal from the chipless tag have been presented. The application of these detection methods for tag identification improves the reader's performance with improved bit error rate. Chapter 5 focuses on noise

Chipless Radio Frequency Identification Reader Signal Processing, First Edition.
Nemai Chandra Karmakar, Prasanna Kalansuriya, Rubayet E. Azim and Randika Koswatta.

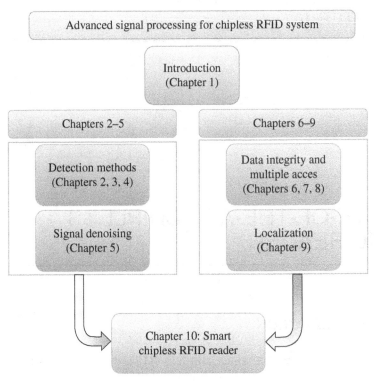

Figure 10.1   Organization of the chapters in the book.

filtering approaches from the backscattered signal to enhance the signal-to-noise ratio of the backscattered signal. Chapter 6 reviews established anticollision algorithms used in conventional RFID systems. However, these algorithms cannot be used in chipless RFID. Therefore, two dedicated anticollision algorithms, (i) *time–frequency* analysis and (ii) FMCW RADAR method, have been proposed in the following two chapters. Chapter 7 covers the *time–frequency* analysis method for collision detection. The comprehensive analysis on FMCW RADAR technique has been presented in Chapter 8. In the end, the concluding Chapter 10 proposes a state-of-the-art smart chipless RFID reader where all the challenges for obtaining a reliable system have been considered.

The chapter organization is shown in Figure 10.2. Section 10.1 is the introductory section. The challenges and various signal processing aspects of the reader are covered in Section 10.2. Section 10.3 presents the proposed smart chipless RFID reader with advanced level signal processing algorithms implemented in the digital section. It includes a

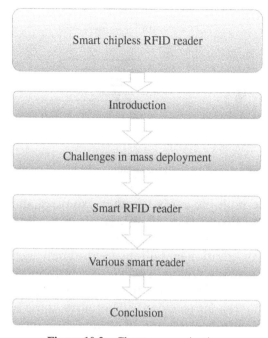

Figure 10.2 Chapter organization.

comprehensive description and discussion of different subsystems of the reader. A flowchart of the working principle of the smart reader is also included. Some application areas of smart reader have been evaluated in Section 10.4 followed by conclusion in Section 10.5.

10.2 CHALLENGES IN MASS DEPLOYMENT

Chipless RFID systems are moving their footsteps into enormous emerging application areas. The mass deployment of chipless RFID systems involves new set of challenges. A reader proficient of addressing the new challenges with improved detection and anticollision algorithms is referred to as *smart chipless RFID reader*. However, the requirements from the reader and associated challenges are application specific. A few examples will help to identify the associated challenges and issues in different applications. In slot reader, as shown in Figure 10.3a, the card with chipless tag is swiped along the slot of the reader. This is the simplest reader with minimal challenges. The main issue is the movement of the card. So the reader needs to be able to read the tag while it's on move. Anticollision algorithm is not required as it's

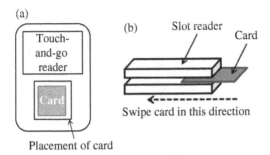

Figure 10.3 (a) Touch-and-go reading. (b) Slot-type reading.

unlikely to swipe multiple cards together. Another application is the touch-and-go-type reader (Fig. 10.3b). This sort of systems is used in authentication of traveling card and person's entry in building and so on. Here, the tag orientation and tilting are the challenges. While touching the card, the tag may not be in proper orientation. As the cards are kept in pocket or holders, they might get tilted, which makes it challenging for the reader to identify. Therefore, these aspects need to be considered in advance. Figure 10.4a and b shows smart shelves and RFID-enabled retail stores, respectively. These are more complex application fields with several significant challenges. In RFID-enabled smart shelf, the tagged books are arranged side by side. While reading with a handheld reader, the antenna may energize multiple tags, which lead to collision. Therefore, anticollision algorithm is required here. Moreover, the orientation issue need also be addressed while designing the reader. A more complex scenario is shown in Figure 10.4b where RFID can be used for authentication, item level tagging, and protection against shoplifting. Therefore, different reader setups are required like handheld reader (for maintaining the stock), static reader (for checkout, antitheft protection), and touch-and-go-type reader (for checkout). Collision scenario is more common and acute here compared to the smart shelves.

The various challenges that are faced by a smart chipless RFID reader are shown in Figure 10.5. A significant amount of research is required to address these challenges. A successful accomplishment of the challenges will make a commercially viable smart chipless RFID reader.

Tag Sensing: Before starting the reading and identification procedure, the reader needs to sense the presence or entrance of the tagged items in the interrogation zone so that the reader can automatically start the reading process without any human operator. After sensing the tag, the reader starts the reading process.

(a)

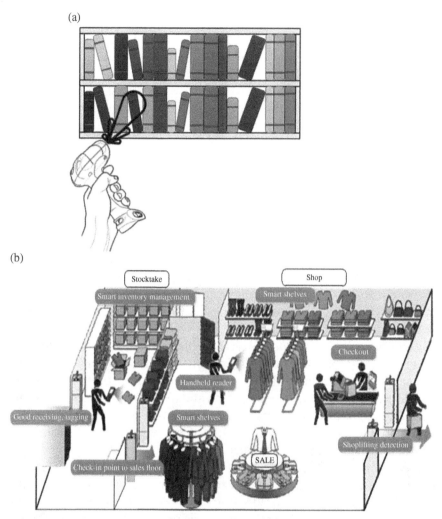

(b)

Figure 10.4   (a) Smart shelve and (b) RFID in retail store.

Collision Detection and Multi-Tag Identification: Simultaneous response from multiple tags may lead to a wrong identification. To avoid the scenario, collision detection is necessary. The reader should be capable of distinguishing between a collision and no-collision scenario. Multi-tag identification is a serious issue in any RFID system. In conventional RFID systems, tag has the capability of signal processing and establishing a two-way communication with the reader. However, in chipless RFID, a two-way communication is not possible, and the tag cannot sense the presence of nearby tags. Therefore, the

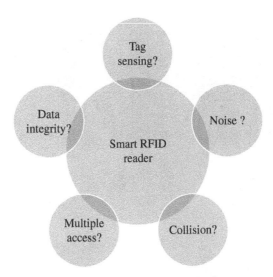

Figure 10.5   Challenges in smart chipless RFID reader.

multi-tag identification problem needs to be solved solely from the reader's side.

Data Integrity and Authenticity: Data integrity means maintaining the accuracy and reliability of the ID extracted from the tag. For keeping the satisfactory accuracy level of the identification of chipless RFID tags, decoded tag IDs need to be validated by error detecting and correction methods. This ensures the system's reliability by detecting errors in the decoded tag IDs. Otherwise, wrong identification may degrade the system's performance.

Denoising and Detection: Chipless RFID identification is affected by noise present in the interrogation zone. The noise need to be mitigated for reliable and accurate identification. Postprocessing on the tag response signal by the detection methods proposed in Chapters 2 and 3 will lead to improved performance of the chipless RFID reader.

The aim of this chapter is to provide a vision for a smart chipless RFID reader that can be used to address all the challenges listed above. This reader is an integral system of all the subsystems and algorithms that has been discussed so far.

10.3   SMART RFID READER

The proposed smart RFID reader comprises of all the subsystems described and proposed in the previous chapters. The proposed reader gives a comprehensive solution of various challenges faced in chipless

RFID system. Figure 10.6 shows the complete block diagram of the proposed smart reader for chipless RFID tag reading. The entire reader is subdivided into two main blocks: (i) IT layer (*front end*) and (ii) physical layer (*back end*). A detailed description of the blocks together with individual subsystems is described in the following sections.

10.3.1 Physical Layer (*Front End*)

As shown in Figure 10.6, the front end contains the reader antenna, RF board, and digital board for chipless RFID reader. Descriptions about each of the subsystems are as follows.

Antenna: Antennas are the integral part for both frequency-domain and time-domain readers. Antennas establish the communication with the remotely placed tags for identification. For the frequency-domain chipless RFID tags, UWB antennas having a wide frequency bandwidth are required. The operating frequency and bandwidth of the reader

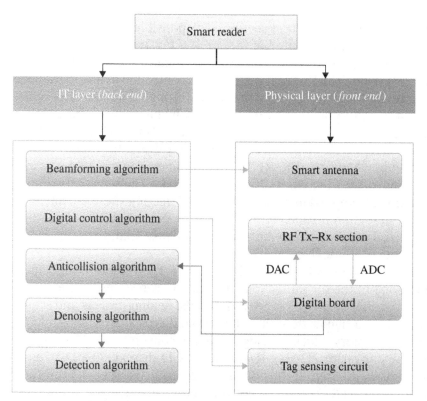

Figure 10.6 Overall block diagram of the proposed smart reader.

antenna depends on the operating frequency of the chipless tag [2, 3]. The antennas may be omnidirectional or wide-beamwidth antennas to cover as much area as possible. However, in some cases, narrow-beam, high-gain antennas are also employed for mitigating the interference from nearby other tags. Instead of using fixed-beam antennas, the smart antenna with beam steering capability is a better option in some scenarios for chipless RFID tag reading. It enables the reader to focus toward a particular tag while mitigating interference from other directions. By steering the beam toward each tag in the interrogation zone, multiple tag identification can be done. Beamforming network is an integral part for the smart antenna. It can be analog or digital depending on the requirement of the system. For short-range communication (<10 cm), 8 dBi gain may be satisfactory, but for long-range communication, a gain greater than 22 dBi is preferable [1].

RF Board: The RF board mainly performs the interrogation of the tag, receiving the response from the tag, and demodulation of the response signals for information extraction. The RF board can be subdivided into three main parts: (i) interrogation signal generation, (ii) communication with tag, and (iii) demodulation of response signal. The RF signal generation section is responsible for generation of the RF signal for interrogation of the chipless tag. It may consist of voltage-controlled oscillator (VCO), frequency mixer, filters (LPF, BPF, HPF, and BSF), etc. The VCO input voltage is controlled by the digital board. The reader communicates with the tag through antennas. The received signal from the tag is then processed. It includes gain–phase detector, filters, and interface with the digital section (Fig. 10.7). The individual components with their specifications and evaluation have been discussed in Chapter 8. The RF board is designed on a high-performance PCB substrate (Taconic TLX-0, 0.5 mm thick). The sensitivity of the RF board is an important issue in reliably detecting the resonances from the backscattered signal from the tag. Improved sensitivity results in improved reading range but may also cause undesired interference from other RF devices [1].

Digital Board: All the control functions for the reader is implemented and done in the digital section. This section is responsible for the VCO control voltage generation, activation of the reading process, and acquiring the response signals from the RF section. It typically contains microprocessor, memory block, analog-to-digital converter (ADC), digital-to-analog converter (DAC), RAM, and LCD display. The DAC is used to convert the digital output to analog to drive the VCO in the RF board. The ADC is used to convert the received signal from tag to a

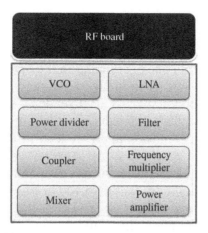

Figure 10.7 Components in RF board.

sampled digital signal. A detailed description about the digital section design, implementation, and evaluation results is presented in Ref. 1.

10.3.2 IT Layer (*Back End*)

Several challenges need to be addressed for full commercialization of the chipless RFID systems. Most of the challenges are faced by the reader as this is the only part among the reader and tag where signal processing and anticollision algorithms can be employed. The front end of the reader remains the same with antennas, digital board, and RF board. The back end can be modified for signal processing algorithms for collision detection, noise cancellation, and improved detection techniques. The back end can actually be merged with the digital section when the algorithms are implemented in FPGA. However, as we are proposing this new dimension in the reader with smart signal processing algorithms, we are presenting this section separately here. The detailed description of several subsystems of the back end is presented in the following paragraphs.

Tag Presence Sensing: For starting the tag reading process, either a human operator is required to start the reading when a tag comes to the interrogation zone or the reader needs to be activated all the time. To avoid both of the hurdles, the reader maybe integrated with sensors to sense the presence of tag in the interrogation zone. As soon as the reader senses the presence of a tag, it can activate the reading process. A detailed description of sensor-based tag presence detection has been discussed in Ref. 1. Using automatic tag presence detection method enables to deploy a fully automated chipless RFID system without the requirement of any human involvement.

Anticollision Algorithm: Collision is an inherent problem in RFID systems. When more than one tag remains in the reading zone, they tend to reply back simultaneously to the reader's query, which gives rise to collision. The developed chipless RFID reader described so far does not use anticollision/collision detection algorithm [4]. Therefore, it is unable to sense the tag collision and may lead to wrong identification due to collision.

The first step for multiple access is to sense and detect the collision and to estimate the number of tags involved in collision. In Chapter 6, a method based on *Linear Block Code* (LBC) has been proposed for collision detection. It requires little modification of the tag resonators, which can be done while designing the tags. From the reader part, after decoding the ID is validated on the basis of the LBC encoded within the tag to find collision. The data encoding with LBC and validation method are described in Chapter 6. Another method of collision detection has been described in Chapter 8 based on FMCW RADAR technique. It requires an FFT on the sampled IF signal coming from the mixer output of the RF board. The FFT algorithm can be implemented on FPGA for collision detection. This method is also capable of estimating the range of each colliding tags. After collision detection, the reader can proceed for individual tag identification if possible through either *time–frequency* analysis or FMCW RADAR technique as shown in Figure 10.8 [5].

Noise Mitigation: The noise sources of the chipless RFID systems can be divided into three main groups: thermal noise introduced from the receiver side, environmental noise (multipath, clutter reflections), and

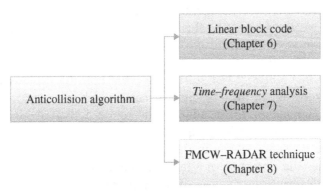

Figure 10.8   Proposed anticollision algorithms for chipless RFID.

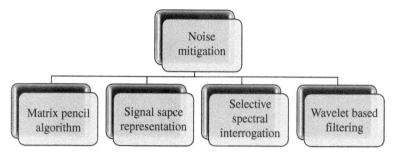

Figure 10.9 Different noise mitigation techniques.

due to fabrication error. The effect of the combined noise is distorted backscattered signal from the tag. Several noise filtering methods have been discussed in Chapters 2 and 3 and shown in Figure 10.9.

The wavelet-based filtering has been tested for time-coded UWB RFID tag with impulse radar-based reading technique. It shows satisfactory performance. This filtering method can be used with time-domain chipless RFID reader for noise cancellation and improved performance [6]. The prolate spheroidal wave function (PSWF) filtering method has been used for reducing the noise effect from the group delay. Using this method, a smooth signal has been reconstructed from the noisy response signals, and resonances have been identified from the phase response of the tag [7]. The matrix pencil algorithm also works well for noise reduction from the backscattered RCS of the tag. It extracts the poles and zeros corresponding to resonances at resonant frequencies [8]. The signal space representation presented in Chapter 2 is an excellent tool for tag ID decoding from noisy response signal. It has been established that in additive white Gaussian noise environment, this method outperforms the fixed threshold-based method for tag identification [9, 10]. Any of these noise mitigation methods is incorporated within the back end of the reader to mitigate the effect of noise in the received response signal from the tag.

Detection Methods: Three improved detection methods have been discussed and presented in Ref. 1, and it introduces the concepts of signal space representation of frequency signatures, time-domain analysis of tag backscatter, and the singularity expansion method as shown in Figure 10.10. Instead of using direct decoding through gain–phase detector, the proposed detection methods will be incorporated with reader to have more reliable identification. After decoding the tag IDs

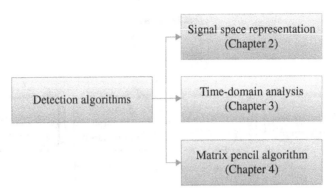

Figure 10.10 Detection methods.

through these methods (gain–phase detector and detection methods), the decoded tag ID can be cross-matched in the back end of the reader to improve the reliability and authenticity.

Data Integrity and Authenticity: Data integrity is an important concern in any communication systems. However, in chipless RFID systems, the tag ID is encoded with 1:1 correspondence. No test bits or check bits are incorporated for data integrity. In Chapter 6, a method using LBC has been introduced for improving the data integrity in chipless RFID systems. It compares the decoded data bits on the basis of LBC to validate the tag ID. In this way, the reader can identify whether the decoded tag ID is correct or wrong. It is very important for authentic identification.

The proposed smart RFID reader has several subsystems that require to control and work simultaneously. Therefore, the working algorithm has several steps.

Figure 10.11 shows a flowchart of the smart identification method. The working flow diagram has several parts. They are discussed in brief in this section. The reading process starts by sensing the presence or entrance of tagged item in the reader's interrogation zone. On sensing the tag's presence, the reader sends the interrogation/query signal toward the interrogation zone. The tags in the reading zone receive the signal and backscatter toward the reader. This part is controlled and done by the front end of the reader (antenna, RF board, digital board, and beamforming network). The next phase is to analyze the received signal for tag ID extraction. This part is carried out in the signal processing back end of the reader. It is already discussed in the previous chapters that the backscattered signal from the tag is analyzed for

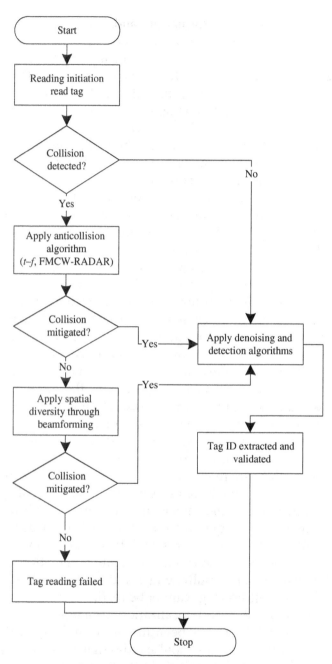

Figure 10.11 Flowchart of the smart chipless RFID reader.

extracting the resonance information and thus the tag ID. When more than one tag simultaneously responds back to the reader, the backscattered signals overlap in time domain, and hence a collided signal is received by the reader. In the proposed smart RFID reader, the received signal is first analyzed with collision detection algorithm for detecting the presence of multiple tags. If there is no collision, then the signal is postprocessed for noise reduction and then the tag ID is decoded with the efficient detection methods like signal space representation and time-domain analysis. Afterward, the decoded tag ID is validated on the basis of LBC. This ensures data integrity and in some cases bit error can also be corrected to some extent.

When collision is detected through the collision detection algorithm, the direct application of the detection method does not work as it processes the collided signal for extraction of a single tag ID. Therefore, the detection method needs to perform anticollision/collision resolution processing for multi-tag identification. The FMCW Radar-based multi-tag identification algorithm is capable of estimating the number of tags colliding when the tags maintain a minimum distance among them, which is greater than the minimum range resolution. It is also capable of extracting the range information of each tag. As shown in Figure 10.12, the range of Tags 1 and 3 can be estimated (proximity tags) but not for Tags 2 and 5 (overlapping tags) as the distance between them is less than the minimum range resolution of the system. After detecting the number of colliding tags and their range, the next step is to identify whether they can be separated through steering the beam of the smart antenna toward one particular tag at a time or any signal separation algorithms are required. As can be seen in Figure 10.12, Tag 1 and Tag 4 can be separately read using beam steering of the smart antenna. However, Tags 1 and 3 fall within the same beam. Thus, a signal separation algorithm is required to separate their individual response signals from the collided response signal. If the separation among the tags in the interrogation zone is not sufficient enough, the reader may fail to individually identify multiple tags and send a notification to the server that due to collision tags cannot be read.

Tag Localization: A novel localization technique for frequency-domain chipless RFID tag has been discussed in Chapter 9 where a short-duration ultra-wideband impulse radio signal (UWB-IR) interrogates the tags and multiple receivers in the interrogation zone capture the backscattered signal from the tags. The received signals from the chipless tags are analyzed for the structural mode radar cross section (RCS) to determine the relative ranges. Using the range information,

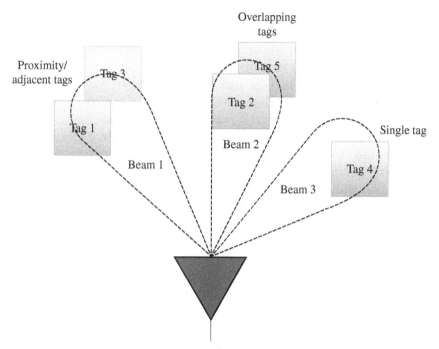

Figure 10.12   Multi-tag scenario in chipless RFID system under spatial diversity.

Linear Least Square (LLS) method is employed for accurate localization of tagged items. The range and angular resolution are 2.1 cm and 3.5°, respectively. The analysis and results create a strong foundation for chipless RFID tags to be used in tracking and localization. Such location-finding method opens up numerous new application areas for chipless RFID systems.

10.4 VARIOUS SMART READERS

The reader's specification for chipless RFID systems depends largely on the type of application as already described in Section 9.2. A more detailed discussion will be presented in this section. Chipless RFID has the potential to be adopted in different application areas like item level tagging, slot card, touch-and-go-type systems, retail, smart library, airport luggage tracking and handling, and vehicle entry system. Different application needs different types of reader. Broadly, they can be classified as two main types as handheld reader and fixed reader.

Handheld reader: The prime concern of handheld reader is the weight and ease of handling. A lightweight, easily portable reader is preferable when using as a handheld reader. The size of the reader is another important aspect. It should not be a big one. Therefore, all the components of smart RFID reader should be incorporated within small RF board, digital board, and FPGA. The power supply should also be included within the reader. The cost of such reader is an important challenge to be dealt with. In general, handheld readers are required in multiple numbers. Therefore, they are preferable to be low cost so they can be bought in numbers without worrying about the cost a lot. As someone needs to operate the handheld reader, it is not necessary to have the automatic tag sensing provision. They can normally be used for short-range identification (Fig. 10.13).

Fixed reader: Fixed readers can be used in vehicle entry and tracking system, baggage tracking, and conveyer belt-type applications. This type of readers are required in smaller number and installed at fixed places in the interrogation zone. Fixed readers are used in more complex reading environment than the handheld reader. An example is the automatic vehicle, person, or item entry monitoring (Fig. 10.14). For vehicle entry tracking in a specified zone, multiple fixed readers are installed at different places. The chipless tags are attached with the cars entering there. For this application, fixed readers are preferable than handheld readers. In this type of applications, the back end signal processing needs to be strong enough to deal with surrounding noise and collision. However, as the weight and size of the reader are not a prime concern here, it needs not to be as small as the handheld one. It can also operate from external power supply, so it doesn't need to have its own power supply.

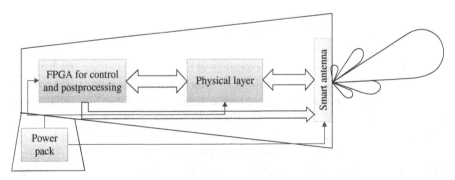

Figure 10.13 Handheld reader for chipless RFID tag reading.

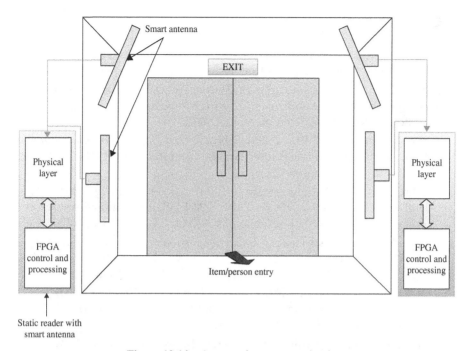

Figure 10.14   Automatic entry monitoring.

10.5 CONCLUSION

Chipless RFID is a low-cost RFID solution that enjoys the advantages of RFID systems but within a fraction of cost. The two main parts in the chipless RFID system are the tag and the reader. However, the tag is a fully passive element, and there is no provision of implementation of signal processing, coding, or modulation scheme within it. Therefore, the key burden lies on the reader part. This book was aimed to focus on the development and research activities related to chipless RFID reader. This book covers the recent achievements in advanced level signal processing methods used and applied in chipless RFID systems for efficient decoding, detection, identification, and localization. Efficient detection methods together with noise mitigation techniques have also been discussed. Afterward, anticollision algorithms for chipless RFID systems have been presented. In this final chapter, an integration of all the subsystems together with the proposed algorithms has been presented as a proposed smart chipless RFID reader. Altogether, this book can be a self-governing solution for chipless RFID reader development methods and challenges to the readers.

REFERENCES

1. N. C. Karmakar, *Chipless RFID Reader Architecture* N. C. Karmakar, R. V. Koswatta, P. Kalansuriya and R. E-Azim, ed. Boston: Artech House, 2013.
2. N. C. Karmakar, "Recent Paradigm Shift in RFID and Smart Antennas," in *Handbook of Smart Antennas for RFID Systems*, N. C. Karmakar, Ed., Hoboken, NJ: John Wiley & Sons, Inc., 2010.
3. N. C. Karmakar, S. M. Roy, and M. S. Ikram, "*Development of smart antenna for RFID reader*," in IEEE International Conference on RFID, Las Vegas, NV, April 16, 2008, pp. 65–73.
4. S. Preradovic and N. C. Karmakar, "*Multiresonator based chipless RFID tag and dedicated RFID reader*," in IEEE MTT-S International Microwave Symposium Digest (MTT), Anaheim, CA, May 23–28, 2010, pp. 1520–1523.
5. R. E. Azim and N. Karmakar, "*Efficient collision detection methods in chipless RFID systems*," presented at the International Conference in Electrical and Computer Systems Engineering (ICECE), Dhaka, Bangladesh, 2012.
6. A. Lazaro, A. Ramos, D. Girbau, and R. Villarino, "Chipless UWB RFID Tag Detection Using Continuous Wavelet Transform," *IEEE Antennas and Wireless Propagation Letters*, 10, pp. 520–523, 2011.
7. W. Dullaert, L. Reichardt, and H. Rogier, "Improved Detection Scheme for Chipless RFIDs Using Prolate Spheroidal Wave Function-Based Noise Filtering," *IEEE Antennas and Wireless Propagation Letters*, 10, pp. 472–5, 2011.
8. M. Manteghi, "*A novel approach to improve noise reduction in the matrix pencil algorithm for chipless RFID tag detection*," in IEEE International Symposium on Antennas and Propagation and CNC-USNC/URSI Radio Science Meeting—Leading the Wave, AP-S/URSI 2010, July 11–17, 2010, Toronto, ON, 2010.
9. P. Kalansuriya, N. Karmakar, and E. Viterbo, in *Chipless and Conventional Radio Frequency Identification: Systems for Ubiquitous Tagging*, N. Karmakar, Ed., Hoboken, NJ: IGI Global, 2012, pp. 218–233.
10. P. Kalansuriya, N. Karmakar, and E. Viterbo, "*Signal space representation of chipless RFID tag frequency signatures*," in 54th Annual IEEE Global Telecommunications Conference: "Energizing Global Communications", GLOBECOM 2011, December 5–9, 2011, Houston, TX, 2011.

INDEX

Chipless Radio Frequency Identification Reader Signal Processing, First Edition.
Nemai Chandra Karmakar, Prasanna Kalansuriya, Rubayet E. Azim and Randika Koswatta.
© 2016 John Wiley & Sons, Inc. Published 2016 by John Wiley & Sons, Inc.

WILEY SERIES IN MICROWAVE AND OPTICAL ENGINEERING

Kai Chang, Series Editor

Texas A&M University

FIBER-OPTIC COMMUNICATION SYSTEMS, Fourth Edition • *Govind P. Agrawal*

ASYMMETRIC PASSIVE COMPONENTS IN MICROWAVE INTEGRATED CIRCUITS • *Hee-Ran Ahn*

COHERENT OPTICAL COMMUNICATIONS SYSTEMS • *Silvello Betti, Giancarlo De Marchis, and Eugenio Iannone*

PHASED ARRAY ANTENNAS: FLOQUET ANALYSIS, SYNTHESIS, BFNs, AND ACTIVE ARRAY SYSTEMS • *Arun K. Bhattacharyya*

HIGH-FREQUENCY ELECTROMAGNETIC TECHNIQUES: RECENT ADVANCES AND APPLICATIONS • *Asoke K. Bhattacharyya*

RADIO PROPAGATION AND ADAPTIVE ANTENNAS FOR WIRELESS COMMUNICATION LINK NETWORKS: TERRESTRIAL, ATMOSPHERIC, AND IONOSPHERIC, Second Edition • *Nathan Blaunstein and Christos G. Christodoulou*

COMPUTATIONAL METHODS FOR ELECTROMAGNETICS AND MICROWAVES • *Richard C. Booton, Jr.*

ELECTROMAGNETIC SHIELDING • *Salvatore Celozzi, Rodolfo Araneo, and Giampiero Lovat*

MICROWAVE RING CIRCUITS AND ANTENNAS • *Kai Chang*

MICROWAVE SOLID-STATE CIRCUITS AND APPLICATIONS • *Kai Chang*

RF AND MICROWAVE WIRELESS SYSTEMS • *Kai Chang*

RF AND MICROWAVE CIRCUIT AND COMPONENT DESIGN FOR WIRELESS SYSTEMS • *Kai Chang, Inder Bahl, and Vijay Nair*

MICROWAVE RING CIRCUITS AND RELATED STRUCTURES, Second Edition • *Kai Chang and Lung-Hwa Hsieh*

MULTIRESOLUTION TIME DOMAIN SCHEME FOR ELECTROMAGNETIC ENGINEERING • *Yinchao Chen, Qunsheng Cao, and Raj Mittra*

DIODE LASERS AND PHOTONIC INTEGRATED CIRCUITS,
Second Edition • *Larry Coldren, Scott Corzine, and Milan Masanovic*

EM DETECTION OF CONCEALED TARGETS • *David J. Daniels*

RADIO FREQUENCY CIRCUIT DESIGN • *W. Alan Davis and Krishna Agarwal*

RADIO FREQUENCY CIRCUIT DESIGN, Second Edition • *W. Alan Davis*

FUNDAMENTALS OF OPTICAL FIBER SENSORS • *Zujie Fang, Ken K. Chin, Ronghui Qu, and Haiwen Cai*

MULTICONDUCTOR TRANSMISSION-LINE STRUCTURES: MODAL ANALYSIS TECHNIQUES • *J. A. Brandão Faria*

PHASED ARRAY-BASED SYSTEMS AND APPLICATIONS • *Nick Fourikis*

SOLAR CELLS AND THEIR APPLICATIONS, Second Edition • *Lewis M. Fraas and Larry D. Partain*

FUNDAMENTALS OF MICROWAVE TRANSMISSION LINES • *Jon C. Freeman*

OPTICAL SEMICONDUCTOR DEVICES • *Mitsuo Fukuda*

MICROSTRIP CIRCUITS • *Fred Gardiol*

HIGH-SPEED VLSI INTERCONNECTIONS, Second Edition • *Ashok K. Goel*

FUNDAMENTALS OF WAVELETS: THEORY, ALGORITHMS, AND APPLICATIONS, Second Edition • *Jaideva C. Goswami and Andrew K. Chan*

HIGH-FREQUENCY ANALOG INTEGRATED CIRCUIT DESIGN • *Ravender Goyal (ed.)*

RF AND MICROWAVE TRANSMITTER DESIGN • *Andrei Grebennikov*

ANALYSIS AND DESIGN OF INTEGRATED CIRCUIT ANTENNA MODULES • *K. C. Gupta and Peter S. Hall*

PHASED ARRAY ANTENNAS, Second Edition • *R. C. Hansen*

STRIPLINE CIRCULATORS • *Joseph Helszajn*

THE STRIPLINE CIRCULATOR: THEORY AND PRACTICE • *Joseph Helszajn*

LOCALIZED WAVES • *Hugo E. Hernández-Figueroa, Michel Zamboni-Rached, and Erasmo Recami (eds.)*

MICROSTRIP FILTERS FOR RF/MICROWAVE APPLICATIONS,
Second Edition • *Jia-Sheng Hong*

MICROWAVE APPROACH TO HIGHLY IRREGULAR FIBER OPTICS
• *Huang Hung-Chia*

NONLINEAR OPTICAL COMMUNICATION NETWORKS • *Eugenio
Iannone, Francesco Matera, Antonio Mecozzi, and Marina Settembre*

FINITE ELEMENT SOFTWARE FOR MICROWAVE ENGINEERING
• *Tatsuo Itoh, Giuseppe Pelosi, and Peter P. Silvester (eds.)*

INFRARED TECHNOLOGY: APPLICATIONS TO ELECTROOPTICS,
PHOTONIC DEVICES, AND SENSORS • *A. R. Jha*

SUPERCONDUCTOR TECHNOLOGY: APPLICATIONS TO
MICROWAVE, ELECTRO-OPTICS, ELECTRICAL MACHINES, AND
PROPULSION SYSTEMS • *A. R. Jha*

TIME AND FREQUENCY DOMAIN SOLUTIONS OF EM PROBLEMS
USING INTEGTRAL EQUATIONS AND A HYBRID
METHODOLOGY • *B. H. Jung, T. K. Sarkar, S. W. Ting, Y. Zhang, Z. Mei, Z.
Ji, M. Yuan, A. De, M. Salazar-Palma, and S. M. Rao*

OPTICAL COMPUTING: AN INTRODUCTION • *M. A. Karim and A. S. S.
Awwal*

INTRODUCTION TO ELECTROMAGNETIC AND MICROWAVE
ENGINEERING • *Paul R. Karmel, Gabriel D. Colef, and Raymond L.
Camisa*

MILLIMETER WAVE OPTICAL DIELECTRIC INTEGRATED
GUIDES AND CIRCUITS • *Shiban K. Koul*

ADVANCED INTEGRATED COMMUNICATION MICROSYSTEMS
• *Joy Laskar, Sudipto Chakraborty, Manos Tentzeris, Franklin Bien, and
Anh-Vu Pham*

MICROWAVE DEVICES, CIRCUITS AND THEIR INTERACTION
• *Charles A. Lee and G. Conrad Dalman*

ADVANCES IN MICROSTRIP AND PRINTED ANTENNAS • *Kai-Fong
Lee and Wei Chen (eds.)*

SPHEROIDAL WAVE FUNCTIONS IN ELECTROMAGNETIC
THEORY • *Le-Wei Li, Xiao-Kang Kang, and Mook-Seng Leong*

MICROWAVE NONCONTACT MOTION SENSING AND ANALYSIS
• *Changzhi Li and Jenshan Lin*

COMPACT MULTIFUNCTIONAL ANTENNAS FOR WIRELESS
SYSTEMS • *Eng Hock Lim and Kwok Wa Leung*

ARITHMETIC AND LOGIC IN COMPUTER SYSTEMS • *Mi Lu*

OPTICAL FILTER DESIGN AND ANALYSIS: A SIGNAL PROCESSING APPROACH • *Christi K. Madsen and Jian H. Zhao*

THEORY AND PRACTICE OF INFRARED TECHNOLOGY FOR NONDESTRUCTIVE TESTING • *Xavier P. V. Maldague*

METAMATERIALS WITH NEGATIVE PARAMETERS: THEORY, DESIGN, AND MICROWAVE APPLICATIONS • *Ricardo Marqués, Ferran Martín, and Mario Sorolla*

OPTOELECTRONIC PACKAGING • *A. R. Mickelson, N. R. Basavanhally, and Y. C. Lee (eds.)*

OPTICAL CHARACTER RECOGNITION • *Shunji Mori, Hirobumi Nishida, and Hiromitsu Yamada*

ANTENNAS FOR RADAR AND COMMUNICATIONS: A POLARIMETRIC APPROACH • *Harold Mott*

INTEGRATED ACTIVE ANTENNAS AND SPATIAL POWER COMBINING • *Julio A. Navarro and Kai Chang*

ANALYSIS METHODS FOR RF, MICROWAVE, AND MILLIMETER-WAVE PLANAR TRANSMISSION LINE STRUCTURES • *Cam Nguyen*

LASER DIODES AND THEIR APPLICATIONS TO COMMUNICATIONS AND INFORMATION PROCESSING • *Takahiro Numai*

FREQUENCY CONTROL OF SEMICONDUCTOR LASERS • *Motoichi Ohtsu (ed.)*

INVERSE SYNTHETIC APERTURE RADAR IMAGING WITH MATLAB ALGORITHMS • *Caner Özdemir*

SILICA OPTICAL FIBER TECHNOLOGY FOR DEVICE AND COMPONENTS: DESIGN, FABRICATION, AND INTERNATIONAL STANDARDS • *Un-Chul Paek and Kyunghwan Oh*

WAVELETS IN ELECTROMAGNETICS AND DEVICE MODELING • *George W. Pan*

OPTICAL SWITCHING • *Georgios Papadimitriou, Chrisoula Papazoglou, and Andreas S. Pomportsis*

MICROWAVE IMAGING • *Matteo Pastorino*

ANALYSIS OF MULTICONDUCTOR TRANSMISSION LINES • *Clayton R. Paul*

INTRODUCTION TO ELECTROMAGNETIC COMPATIBILITY, Second Edition • *Clayton R. Paul*

ADAPTIVE OPTICS FOR VISION SCIENCE: PRINCIPLES, PRACTICES, DESIGN AND APPLICATIONS • *Jason Porter, Hope Queener, Julianna Lin, Karen Thorn, and Abdul Awwal (eds.)*

ELECTROMAGNETIC OPTIMIZATION BY GENETIC ALGORITHMS • *Yahya Rahmat-Samii and Eric Michielssen (eds.)*

INTRODUCTION TO HIGH-SPEED ELECTRONICS AND OPTOELECTRONICS • *Leonard M. Riaziat*

NEW FRONTIERS IN MEDICAL DEVICE TECHNOLOGY • *Arye Rosen and Harel Rosen (eds.)*

ELECTROMAGNETIC PROPAGATION IN MULTI-MODE RANDOM MEDIA • *Harrison E. Rowe*

ELECTROMAGNETIC PROPAGATION IN ONE-DIMENSIONAL RANDOM MEDIA • *Harrison E. Rowe*

HISTORY OF WIRELESS • *Tapan K. Sarkar, Robert J. Mailloux, Arthur A. Oliner, Magdalena Salazar-Palma, and Dipak L. Sengupta*

PHYSICS OF MULTIANTENNA SYSTEMS AND BROADBAND PROCESSING • *Tapan K. Sarkar, Magdalena Salazar-Palma, and Eric L. Mokole*

SMART ANTENNAS • *Tapan K. Sarkar, Michael C. Wicks, Magdalena Salazar-Palma, and Robert J. Bonneau*

NONLINEAR OPTICS • *E. G. Sauter*

APPLIED ELECTROMAGNETICS AND ELECTROMAGNETIC COMPATIBILITY • *Dipak L. Sengupta and Valdis V. Liepa*

COPLANAR WAVEGUIDE CIRCUITS, COMPONENTS, AND SYSTEMS • *Rainee N. Simons*

ELECTROMAGNETIC FIELDS IN UNCONVENTIONAL MATERIALS AND STRUCTURES • *Onkar N. Singh and Akhlesh Lakhtakia (eds.)*

ANALYSIS AND DESIGN OF AUTONOMOUS MICROWAVE CIRCUITS • *Almudena Suárez*

ELECTRON BEAMS AND MICROWAVE VACUUM ELECTRONICS • *Shulim E. Tsimring*

FUNDAMENTALS OF GLOBAL POSITIONING SYSTEM RECEIVERS: A SOFTWARE APPROACH, Second Edition • *James Bao-yen Tsui*

SUBSURFACE SENSING • *Ahmet S. Turk, A. Koksal Hocaoglu, and Alexey A. Vertiy (eds.)*

RF/MICROWAVE INTERACTION WITH BIOLOGICAL TISSUES • *André Vander Vorst, Arye Rosen, and Youji Kotsuka*

InP-BASED MATERIALS AND DEVICES: PHYSICS AND TECHNOLOGY • *Osamu Wada and Hideki Hasegawa (eds.)*

COMPACT AND BROADBAND MICROSTRIP ANTENNAS • *Kin-Lu Wong*

DESIGN OF NONPLANAR MICROSTRIP ANTENNAS AND TRANSMISSION LINES • *Kin-Lu Wong*

PLANAR ANTENNAS FOR WIRELESS COMMUNICATIONS • *Kin-Lu Wong*

FREQUENCY SELECTIVE SURFACE AND GRID ARRAY • *T. K. Wu (ed.)*

PHOTONIC SENSING: PRINCIPLES AND APPLICATIONS FOR SAFETY AND SECURITY MONITORING • *Gaozhi Xiao and Wojtek J. Bock*

ACTIVE AND QUASI-OPTICAL ARRAYS FOR SOLID-STATE POWER COMBINING • *Robert A. York and Zoya B. Popoviać (eds.)*

OPTICAL SIGNAL PROCESSING, COMPUTING AND NEURAL NETWORKS • *Francis T. S. Yu and Suganda Jutamulia*

ELECTROMAGNETIC SIMULATION TECHNIQUES BASED ON THE FDTD METHOD • *Wenhua Yu, Xiaoling Yang, Yongjun Liu, and Raj Mittra*

SiGe, GaAs, AND InP HETEROJUNCTION BIPOLAR TRANSISTORS • *Jiann Yuan*

PARALLEL SOLUTION OF INTEGRAL EQUATION-BASED EM PROBLEMS • *Yu Zhang and Tapan K. Sarkar*

ELECTRODYNAMICS OF SOLIDS AND MICROWAVE SUPERCONDUCTIVITY • *Shu-Ang Zhou*

MICROWAVE BANDPASS FILTERS FOR WIDEBAND COMMUNICATIONS • *Lei Zhu, Sheng Sun, and Rui Li*

FUNDAMENTALS OF MICROWAVE PHOTONICS • *Vincent Jude Urick Jr., Jason Dwight McKinney, and Keith Jake Williams*

RADIO-FREQUENCY INTEGRATED-CIRCUIT ENGINEERING • *Cam Nguyen*

ARTIFICIAL TRANSMISSION LINES FOR RF AND MICROWAVE APPLICATIONS • *Ferran Martín*

PASSIVE MACROMODELING • *Stefano Grivet-Talocia and Bjørn Gustavsen*

Printed in the USA
K019451SCI030916   01S29053000000001050